SpringerBriefs in Electrical and Computer Engineering

Cooperating Objects

Series editor

Pedro José Marrón, Duisburg, Germany

For further volumes:
http://www.springer.com/series/10208

Stamatis Karnouskos · Pedro José Marrón
Giancarlo Fortino · Luca Mottola
José Ramiro Martínez-de Dios

Applications and Markets for Cooperating Objects

 Springer

Stamatis Karnouskos
SAP
Karlsruhe
Germany

Pedro José Marrón
University of Duisburg-Essen
Duisburg
Germany

Giancarlo Fortino
University of Calabria
Calabria
Italy

Luca Mottola
Dipartimento di Elettronica ed Informazione
Politecnico di Milano
Milano
Italy

José Ramiro Martínez-de Dios
University of Seville
Seville
Spain

ISSN 2191-8112   ISSN 2191-8120   (electronic)
ISBN 978-3-642-45400-4   ISBN 978-3-642-45401-1   (eBook)
DOI 10.1007/978-3-642-45401-1
Springer Heidelberg New York Dordrecht London

Library of Congress Control Number: 2013956342

Printed on acid-free paper

Springer is part of Springer Science+Business Media (www.springer.com)

Preface

The book you have in your hands focuses on the applications and markets for Cooperating Objects. Although the Cooperating Objects domain is relatively new, and shares common ground with other domains such as Internet of Things, Sensor Networks, Ubiquitous Computing, Cyber-Physical Systems and Systems of Systems, we can already clearly distinguish the key aspects of "cooperation" as being the key driver behind several application developments.

The applications depicted in this book constitute some of the examples in this promising new domain, and most of which have also been demonstrated within the Cooperating Objects Network of Excellence (CONET). CONET (www.cooperating-objects.eu) is a European project co-funded by the European Commission with the aim to identify and produce work on the main research topics in Cooperating Objects, thus shaping the academic and industrial research in the short, medium and long-terms.

The hereby included applications depict not only futuristic scenarios, but also hands-on experiences and results. To what extent these applications will pave their way to commercial products in the future is too early to tell. We try to take an overview of the market as such, that could influence the Cooperating Objects domain, or vice versa, that could be impacted by Cooperating Objects.

We hope you enjoy a more practical view under the prism of application development in the domain of Cooperating Objects, and that these example applications may spark some new ideas for future innovative approaches.

Karlsruhe, Germany, January 2013   Stamatis Karnouskos
Duisburg, Germany   Pedro José Marrón
Rende, Italy   Giancarlo Fortino
Milano, Italy   Luca Mottola
Seville, Spain   José Ramiro Martínez-de Dios

Acknowledgments

The following contributors made this book possible:

Nils Aschenbruck	Universität Osnabrück and Fraunhofer FKIE
Jan Bauer	Universität Osnabrück
Davide Brunelli	Università di Bologna
Enrique Casado	Boeing Research and Technology—Europe
Armando Walter Colombo	Schneider Electric and Hochschule Emden/Leer
Gianluca Dini	Università di Pisa
Elisabetta Farella	Università di Bologna
Giancarlo Fortino	Università della Calabria
Christoph Fuchs	Universität Bonn and Fraunhofer FKIE
Roberta Giannantonio	Telecom Italia
Philipp Maria Glatz	TU Delft
Raffaele Gravina	Università della Calabria
Stamatis Karnouskos	SAP
Paulo Leitão	Polytechnic Institute of Bragança
Pedro José Marrón	Universität Duisburg-Essen
J. R. Martínez-de Dios	University of Seville
Simone Martini	Università di Pisa
Marco Mendes	Schneider Electric
Luis Merino	University Pablo de Olavide
Daniel Minder	Universität Duisburg-Essen
Luca Mottola	Politecnico di Milano and SICS Swedish ICT
Amy L. Murphy	Fondazione Bruno Kessler
Aníbal Ollero	University of Seville
Lucia Pallottino	Università di Pisa
Gian Pietro Picco	Università degli Studi di Trento
José Pinto	Universidade do Porto
Thiemo Voigt	SICS Swedish ICT

We would also like to express our gratitude to all of the members of the CONET(www.cooperating-objects.eu) consortium that have contributed greatly with their knowledge, insight into the topics and discussions to improve this book.

Contents

Acronyms

AAO	Autonomous Aircraft Operations
ADS-B	Automatic Dependent Surveillance—Broadcast
ANSP	Air Navigation Server Providers
AOC	Airline Operation Center
ASAS	Airborne Separation Assurance System
ASD	Accumulated State Density
ATC	Air Traffic Control
ATCO	Air Traffic Controllers
ATM	Air Traffic Management
billion	10^9 = one billion (short scale number naming system)
CDM	Collaborative Decision Making
CNS	Communication, Navigation and Surveillance
CO	Cooperating Object
CONET	Cooperating Objects Network of Excellence
COTS	Commercial off-the-shelf
CR	Collaborative Routing
DPWS	Devices Profile for Web Services
FPGA	Field Programmable Gate Array
GPS	Global Positioning System
GUI	Graphical User Interface
ICAO	International Civil Aviation Organization
IFR	Instrument Flight Rules
IP	Internet Protocol
LiFo	Last in First out
MANET	Mobile Ad Hoc Network
million	10^6 = one million (short-scale number naming system)
ODMRP	On-Demand Multicast Routing Protocol
OoS	Out-of-Sequence
ORS	Open Route Service
OSM	OpenStreetMap
pdf	Probability density function
REST	REpresentational State Transfer

RGCS	Review of General Concept of Separation
RSSI	Radio Signal Strength Indication
SESAR	Single European Sky ATM Research
STMP	Sensor Data Transmission and Management Protocol
SUA	Special Use Airspace
SWIM	System Wide Information Management
TBO	Trajectory-Based Operations
TCO	Total Cost of Ownership
trillion	10^{12} = one trillion (short-scale number naming system)
UAN	Underwater Acoustic Network
WLAN	Wireless Local Area Network
WMN	Wireless Mesh Network
WS-DD	OASIS Web Services Discovery and Web Services Devices Profile
WSN	Wireless Sensor Network

Chapter 1
Introduction

1.1 Cooperating Objects Context

The core idea behind amalgamating the physical and virtual (business) world is to seamlessly gather useful information about objects of the physical world and use the information in various applications in order to provide some added value. In the last years we have witnessed a paradigm change, where the rapid advances in computational and communication capabilities of embedded systems, are paving the way towards highly sophisticated networked devices that will be able to carry out a variety of tasks not in a standalone mode as usually done today, but taking into full account dynamic and context specific information, and following dynamic collaborative approaches.

These "objects" will be able to cooperate, share information, act as part of communities and generally be active elements of a more complex system. The close interaction of the business and real world will be achieved by auxiliary services provided in a timely fashion from networked embedded devices. These will be able to collaborate not only among them but also with on-line services, that will enhance their own functionality.

As already defined [1], one can consider the that :

"Cooperating Objects are modular systems of autonomous, heterogeneous devices pursuing a common goal by cooperation in computations and in sensing and/or actuating with the environment."

The domain of Cooperating Objects is a cross-section between (networked) embedded systems, ubiquitous computing and (wireless) sensor networks. There are, therefore, several flavours of Cooperating Objects depending on the degree in which they fulfil different features. Some of them can process the context of cooperation

---

Contributors of this chapter include: Stamatis Karnouskos, Pedro José Marrón, and Daniel Minder.

S. Karnouskos et al., *Applications and Markets for Cooperating Objects*,
SpringerBriefs in Cooperating Objects,
DOI: 10.1007/978-3-642-45401-1_1, © The Author(s) 2014

2 1 Introduction

intentionally, act on it and intentionally extend it, change it or stop it. As such they may possess the necessary logic to understand semantics and build complex behaviours, thus allowing the Cooperating Object to be part of a dynamic complex ecosystem.

1.2 Magna Carta of Featured Applications

Although the domain of Cooperating Objects is an emerging one, there have already been applications that tackle issues dominant in what one would typically characterise as Cooperating Objects applications, systems and services. The aim is to shortly depict some of these applications, in order to provide a glimpse on the hands-on experiences of the vision and its impact. The list of the applications depicted in this book is by no mean complete, and should serve only as an indicator of steps to be followed in order to realise the Cooperating Object vision.

The applications depicted in Table 1.1 serve as indicative examples from various domains that have been selected, which constitute also the "Magna Carta" for

Table 1.1 Overview of featured applications

Category	Section	Application
Deployment and management of coorporating objects	2.2	Monitoring railway bridges
	2.3	Cooperative industrial automation systems
	2.4	Light-weight bird tracking sensor nodes
	2.5	Public safety scenarios
	2.6	Road tunnel monitoring and control
Mobility of cooperating objects	3.2	Mobility in industrial scenarios
	3.3	Mobility in air traffic management
	3.4	Mobility in ocean scenarios
	3.5	Person assistance in urban scenarios
	3.6	Mobility in civil security and protection
Cooperating objects in healthcare applications	4.2	Physical activity recognition
	4.3	Real-time physical energy expenditure
	4.4	Emotional stress detection
	4.5	Physical rehabilitation
	4.6	Energy aware fall detection
	4.7	Distributed digital signal processing
	4.8	Model predictive control

this book. An additional overview with respect to the integration aspects of each application is given in Table 1.2, while a categorization with respect to the design space is shown in Table 1.3.

1.3 Integration Aspects

Achieving enhanced system intelligence by cooperation of smart embedded devices pursuing common goals is relevant in many types of perception and system environments. In general, such devices with embedded intelligence and sensing/actuating capabilities are heterogeneous, yet they need to interact seamlessly and intensively over wired and/or wireless networks. More constrained devices may also cooperate with more powerful (or less congested) neighbours to meet service requests, opportunistically taking advantage of global resources and processing power. Independently of the structuring level (weakly structured or highly structured), process-driven applications make use of different kinds of data resources and combine them to achieve the application task.

Cooperation between objects can be understood in the following context:

- Two (or more) objects (i.e., object-to-object or object-to-business) are able to engage into a conversation in a loosely-coupled manner.
- The objects have a common understanding of well-defined communication patterns and protocols.
- The objects are able to exchange data relevant to their capabilities and needs.
- The objects share computational resources when needed by means of information migration or data mash-ups.
- The objects are able to cluster in order to create distributed data gathering/ processing platforms.

Adding cooperation in this context makes it imperative to have a look from a different angle, i.e., that of integration with the goal of cooperation.

Several aspects might surface when developing and integrating Cooperating Object applications and services [2], some of which are:

- **Dynamic collaboration**: Devices with sensing and/or actuating capabilities and embedded intelligence should be able to dynamically collaborate in the environment and provide services to the user, which can be an end-user, a device or another service. As dynamic collaboration is the foundation for any cooperation, this requirement is frequently tackled by existing middleware developments.
- **Extensibility**: Flexible support for extending the capabilities of a device is needed. Cooperating Objects is a rapidly developing domain and implementation should take future growth into consideration. Since extensions can be made through the addition of new functionality or modification of existing one, support for change should be provided while minimizing impact to existing system functions. One possibility to achieve this is to support protocol composition [3] or to rely on adequate programming abstractions [4, 5].

- **Optimal resource utilization**: Optimal management of resources at the local (device) as well as non-local (groups, global view) level is needed. As most of the Cooperating Objects are expected to be resource-constrained, the resource utilization should be considered and possibly captured in a cooperation context. For instance, it should be possible that resource-scarce devices exploit the capabilities of devices with more resources, and opportunistically take advantage of the resources in the surroundings if it makes sense from the strategy or performance viewpoint. So far, such optimizations have been done only with respect to individual system functions, e.g., [7] or [8].

- **Description of objects (interface)**: In order to enable Cooperating Object interactions, the implementation independent description of the object that can be used by both implementers and requesters must be available. This will facilitate decoupling of design and actual implementation, which will enable cooperating concepts to be developed in a loosely coupled way with respect to the actual software and hardware available. Due to its general usefulness, this requirement is realized by many existing integration systems, e.g., [9] or [10].

- **Semantic description capabilities**: Semantics and ontologies should be used to enforce the dynamic interpretation of things and as an input for reasoning systems. An object should be able to not only understand that cooperation is possible, but also to assess what impact the cooperation might have on the resources, time, processor utilization, among others. Thereby, it should describe constraints of capabilities of the specific cooperation. So far, only few existing integration approaches such as PECES [11] attempt to tackle this requirement. However, it is important to notice that the use of semantic descriptions has been long researched at industrial scale, for example in the manufacturing industry [12, 13], from where it has been possible to obtain experience about its potential application in embedded devices.

- **Inheritance/polymorphism**: To simplify programming via code reuse, it would make sense to have a way to form new objects using objects that have already been defined. At a later stage one can move towards the Composite Reuse Principle, which enables polymorphic behaviour and code reuse by containing other classes that implement the desired functionality. This approach is partially addressed in [4, 14].

- **Composition/orchestration**: As basis for cooperation, generation and execution of work plans between objects, services and other resources should be supported in order to promote their interaction. Examples for this type of orchestration are the adaptable flows implemented in ALLOW [6].

- **Pluggability**: Due to the continuous evolution of future systems, ubiquitous integration will require the dynamic interaction with newly plugged-in and previously unknown objects. This refers not only to software but also to hardware; typical examples include communication, computation, behaviour, etc. and calls for a component-based approach where modules can be combined to customize existing behaviour or to deliver more complex ones. Cooperating Objects supporting pluggability will enable third-party developers to create capabilities to extend them, easy ways of adding new features, reduced size and independent application development, etc. On the software side, this can be addressed by approaches such as Speakeasy [15].

- **Service discovery**: Cooperating Objects must support a mechanism for each node to make its services known to the system and also to allow querying for services. Automatic service discovery will allow us to access them in a dynamic way without having explicit task knowledge and the need of a priori binding. The last would also enable system scalability and support the composable approach of services. However, existing approaches such as [16] or [17] mostly focus on low-level aspects of this process.

- **Service direct device access**: Applications must be able not only to discover but, in many cases, also to communicate directly with devices, and consume the services they offer [18]. The need to bypass intermediates and directly acquire specific data from the device may offer business benefits and rapid development, deployment, and change management. Additional support, e.g., the capability of event notifications from the device side to which other services can subscribe, may provide optimization advantages.

- **Service indirect device access (gateway/mediator)**: Gateways might glue the Cooperating Object infrastructure devices by hiding heterogeneity and resource scarceness. However, most efforts in the research domain today focus on how to open the device functionality to the enterprise systems, yet, the opposite, i.e., the opening of enterprise systems to the devices, might also be beneficial [18]. For instance, devices should be able to subscribe to events and use enterprise services; this can be achieved by creating "virtual devices" that proxy an (enterprise) service. Having achieved that, business logic running locally on devices can now take decisions not only based on its local information, but also on information from enterprise systems.

- **Brokered access to events**: Events are a fundamental pillar of a service-based infrastructure; therefore access to these has to be eased. As many devices are expected to be mobile, and their on-line status often changes (including the services they host), buffered service invocation should be in place to guarantee that any started process will continue when the device becomes available again (opportunistic networking). Also, since not all applications expose (web) services, a pull point should be realized that will offer access to infrastructure events by polling [19]. Minimized resource usage on the device by delegating access to a more powerful device/system will be beneficial.

- **Service life-cycle management**: In future infrastructures, various services are expected to be installed, updated, deleted, started, and stopped. Therefore, the requirement is to provide basic support on-device/in-infrastructure that can offer an open way of handling these issues [20, 21].

- **Legacy device integration**: Devices of older generations should be also part of the new infrastructure. Although their role will be mostly providing (and not consuming) information, we have to make sure that this information can be acquired and transformed [18] to fit in the new service-enabled infrastructure. The latter is expected to be achieved via the wrapping of them, for example, using web services. An alternative is to use extensible protocol composition [22, 23].

- **Historian**: In an information-rich infrastructure, a continuous logging of relevant data, events, and the history of devices is needed. The historian is needed to offer logging of information to services, especially when an analysis of up-to-now behaviour of devices and their services is needed, for example, to support system audits.
- **Device management**: Service-enabled devices will contain both, static and dynamic data. This data can now be better and more reliably integrated, e.g., into enterprise systems. However, in order to manage large infrastructures, a common way of applying basic management tasks is needed [24, 25]. The device management requirement makes sure that at least on the middleware side, there is a way to hide heterogeneity and provide uniform access to a device's and infrastructure's life cycle.
- **Service monitoring**: Anticipating that the overall infrastructure will rely on services, it should be possible to monitor these services and determine their status [18]. Based on their continuous monitoring, key performance indicators can be acquired, e.g., responsiveness, reliability, performance, quality, etc.
- **Security, trust and privacy**: Security, trust and privacy mechanisms should be considered. Access to the devices and their services will depend on the deployed security context and, therefore, basic functions should be supported. Trust relationships will need to be considered and built upon. Similarly, privacy should be preserved especially for devices operating in sensitive user areas, e.g., hospitals, households, etc. This requires the development of new methods or the adaptation of existing ones to new application areas [20, 26].

An overview with respect to the integration aspects of each application is given in Table 1.2.

1.4 Design Space

Cooperating Objects share common ground [1] with several domains such as software agents, Internet of Things, Cyber-Physical Systems, System of Systems, etc. Hence, it is natural that they share also a common design space. However, the distinguishing difference is that the collaboration is playing a pivotal role as well as the cross-layer interaction among different devices, systems, services and applications.

We have already identified several Cooperating Object characteristics which are depicted in more detail in [1]. More specifically we have:

- **Modularity**: A Cooperating Object is composed of several elements that need to exhibit certain features. Each of the elements contributes to the functionality of the overall Cooperating Object, but the modularisation helps to keep the single devices simple and maintainable. The modular design makes it possible to replace an element by a more powerful one or to add new ones that extend the functionality. Thus, the Cooperating Object can be developed in an evolutionary fashion and adapted to new needs.

Table 1.2 Featured applications and their integration aspects

Integration ⇒ / Applications ⇓	Service direct device access	Service indirect device access (gateway/mediator)	Brokered access to events	Service life-cycle management	Legacy device integration	Historian	Device management	Service monitoring	Security, trust and privacy
Monitoring Railway Bridges	○	◐	◐	○	○	●	◐	●	○
Cooperative Industrial Automation Systems	●	●	●	◐	●	●	◐	◐	○
Light-weight Bird Tracking Sensor Nodes	●	●	◐	○	○	●	○	◐	○
Public Safety Scenario	●	●	●	◐	◐	◐	◐	●	◐
Road Tunnel Monitoring and Control	○	◐	◐	○	○	●	◐	●	◐
Mobility in Industrial Scenarios	●	●	●	◐	●	●	●	●	◐
Mobility in Air Traffic Management	●	●	●	◐	●	●	●	●	●
Mobility in Ocean Scenarios	●	●	●	○	◐	●	◐	◐	●
Person Assistance in Urban Scenarios	○	●	●	◐	○	●	◐	◐	◐
Mobility in Civil Security and Protection	○	●	●	◐	◐	●	◐	●	●
Physical Activity Recognition	●	◐	◐	○	◐	●	●	●	●
Real-time Physical Energy Expenditure	●	◐	◐	○	◐	●	●	●	●
Emotional Stress Detection	●	◐	◐	○	◐	●	●	●	●
Physical Rehabilitation	●	◐	◐	○	◐	●	●	●	●
Energy Aware Fall detection	●	◐	◐	○	◐	●	●	●	○
Distributed Digital Signal Processing	◐	●	◐	○	◐	●	●	●	○
Model Predictive Control	●	◐	◐	○	◐	●	●	●	○

Integration ⇒ / Applications ⇓	Dynamic collaboration	Extensibility	Optimal Resource utilisation	Description of objects (interface)	Semantic description capabilities	Inheritance/ polymorphism	Composition/ orchestration	Pluggability	Service discovery
Monitoring Railway Bridges	◐	●	●	○	○	○	○	◐	◐
Cooperative Industrial Automation Systems	●	●	◐	●	◐	◐	●	●	●
Light-weight Bird Tracking Sensor Nodes	◐	●	●	○	○	○	○	○	◐
Public Safety Scenario	◐	●	●	○	○	◐	○	○	◐
Road Tunnel Monitoring and Control	◐	●	●	○	○	○	○	◐	◐
Mobility in Industrial Scenarios	●	●	●	◐	●	●	●	●	●
Mobility in Air Traffic Management	●	●	○	◐	○	○	◐	●	●
Mobility in Ocean Scenarios	●	●	●	◐	○	◐	●	●	○
Person Assistance in Urban Scenarios	●	◐	●	○	○	○	◐	○	○
Mobility in Civil Security and Protection	●	●	●	○	○	◐	●	●	○
Physical Activity Recognition	◐	●	●	●	◐	●	◐	●	●
Real-time Physical Energy Expenditure	◐	●	●	●	◐	●	◐	●	●
Emotional Stress Detection	◐	●	●	●	◐	●	◐	●	●
Physical Rehabilitation	◐	●	●	●	◐	●	◐	●	●
Energy Aware Fall detection	◐	●	●	●	○	○	○	●	○
Distributed Digital Signal Processing	◐	●	●	●	○	○	○	●	○
Model Predictive Control	◐	●	●	●	○	○	○	●	○

Legend: ● Covered, ◐ Partially covered, ○ Not covered

- **Autonomy**: Each Cooperating Object element can decide on its own about its involvement in a Cooperating Object. If the element does not participate at all in the cooperation and coordination activities, it is not considered part of the Cooperating Object. Otherwise, it decides about the degree of participation. In general, an element can dedicate only a fraction of its resources or its functionality to the current Cooperating Object, thus leaving the possibility to serve multiple Cooperating Objects.
- **Heterogeneity**: In Cooperating Objects, heterogeneity is a crucial point since it is more than heterogeneity in terms of, e.g., processing power or memory. In fact, a Cooperating Object must combine devices of different system concepts, i.e., Wireless Sensor Networks, embedded systems, robotics, etc. Since these elements can have different hardware characteristics, heterogeneity is also exhibited as a consequence.
- **Computation**: Due to the different nature of the single CO elements in a Cooperating Object the computational capabilities can vary largely. However, a device must at least be able to take an autonomous decision about its involvement in a Cooperating Object and to communicate with others, which usually requires also computation.
- **Interaction with the environment**: Cooperating Objects interact with the environment using sensors and/or actuators. The interaction with the environment should be substantial, especially with respect to actuators, i.e., actuation should have a changing effect on the environment. The involvement of sensors and actuators makes Cooperating Objects real-world objects, i.e., there are no pure virtual Cooperating Objects. The interaction with the environment must be a core functionality of the Cooperating Object and not just an optional side-effect.
- **Communication**: If a device communicates there are three techniques of information exchange [27]. The most common technique is explicit communication, which can be performed using various means, e.g., wires, radio, light, sound. The content of the communication is manifold and can range from just the state of the single element to a common planning. Besides explicit communication, there are two other techniques that work by observation using sensors. With passive action recognition the actions of other devices are observed, e.g., if an actuator moves. In contrast, the effects of actions of others can be sensed ("stigmergy"), e.g., the increase of temperature caused by a heater. Usually, these forms of communication show the lack of common interfaces for direct communication. Nevertheless, the inclusion of such devices allows for interesting applications.
- **Common goal**: The ultimate reason for a Cooperating Object to exist is the common goal it tries to achieve. There should be a reason for pursuing the goal using Cooperating Objects: either the goal can only be achieved through Cooperating Objects or there is at least an improvement compared to a monolithic or centralised approach. Although the devices do not know the overall goal they execute a task to achieve it. Thus, each device has detailed knowledge only about its area of responsibility, but limited information about the whole Cooperating Object. However, the cooperation of the single devices make it possible to achieve the overall goal,

Table 1.3 Design space of featured applications

Design Space ⇒ Applications ⇓	Modularity	Autonomy	Heterogeneity	Computation	Interaction with the Environment	Communication	Common Goal	Cooperation
Monitoring Railway Bridges	●	○	●	●	●	●	●	●
Cooperative Industrial Automation Systems	●	◐	●	●	◐	◐	●	◐
Light-weight Bird Tracking Sensor Nodes	●	◐	●	◐	●	●	◐	◐
Public Safety Scenarios	●	◐	◐	◐	◐	●	●	●
Road Tunnel Monitoring and Control	●	○	●	●	●	●	●	●
Mobility in Industrial Scenarios	●	●	●	●	●	●	●	●
Mobility in Air Traffic Management	●	◐	●	●	●	●	○	●
Mobility in Ocean Scenarios	●	●	●	◐	●	●	●	●
Person Assistance in Urban Scenarios	●	●	◐	◐	●	●	●	●
Mobility in Civil Security and Protection	●	●	●	◐	●	●	●	●
Physical Activity Recognition	●	○	●	◐	○	●	○	○
Real-time Physical Energy Expenditure	●	○	●	◐	○	●	○	○
Emotional Stress Detection	●	○	●	◐	○	●	○	○
Physical Rehabilitation	●	○	●	◐	○	●	○	○
Energy Aware Fall detection	●	○	●	◐	◐	●	◐	◐
Distributed Digital Signal Processing	●	○	●	●	◐	●	◐	◐
Model Predictive Control	●	○	●	◐	◐	●	◐	◐

Legend: ● Covered, ◐ Partially covered, ○ Not covered

which needs the full picture. Thus, the intelligence of the system lies distributed in the network.

- **Cooperation**: In Cooperating Objects cooperation is always intentional and driven by a goal. Without a goal and, thus, no tasks, there is no need for cooperation at all. Although unintentional interaction might deliver the same results it does not happen in a controlled way which creates problems in case of errors. For example, reconfiguration is more difficult if the exact task that a device has performed is not known. The participation of all devices in a Cooperating Object is needed to achieve the common goal, i.e., a Cooperating Object is more than just the sum of the single devices. Nevertheless, the common goal does not imply benefits for all the cooperating devices. Some of them can be especially designed to help in cooperation, others can play a more active part in one cooperative task to profit more in another one. When autonomous and selfish objects decide autonomously if and how they cooperate, the sum of the benefit must be positive. Otherwise, a device will eventually not agree to cooperate or not be asked to cooperate any more.

A categorization of the featured applications with respect to the design space is shown in Table 1.3.

1.5 Conclusions

The domain of Cooperating Objects is an emerging one, that envisions the wide-spread collaboration between devices, systems and applications in a fully networked world. Existing efforts are seen overwhelmingly in academia and in some cutting edge cooperation with the industry. From an analysis on the design space of the applications also described in this book, we can see that the issues tackled are mostly modularity, heterogeneity, communication, environment interaction and computation, which pose the basis for creating cooperative scenarios. Some aspects such as autonomy, common goal interactions and cooperation still are (as expected) in early stages.

As Cooperating Objects deal with the amalgamation of physical and virtual world, clearly the integration is a key aspect that comes up in all of the approaches. Service-based integration is on the rise, while management, monitoring and extensibility are key targets as well. Most of the approaches also focus on the resource usage of the Cooperating Objects, as these are in their large majority resource-constrained devices (at least in the depicted scenarios). Some efforts are done towards legacy system integration and migration of them with inclusion of new technologies. However, full infrastructure lifecycle management as well as semantic support still lack behind. Security, trust and privacy, are of key importance and although some approaches consider it, the right equilibrium between functionality, application domain and resource management still needs to be found.

References

1. Marrón PJ, Minder D, Karnouskos S (2012) The emerging domain of cooperating objects: definition and concepts. Springer, Berlin. doi:10.1007/978-3-642-28469-4
2. Karnouskos S, Vilaseñor V, Handte M, Marrón PJ (2011) Ubiquitous Integration of Cooperating Objects. Int J Next-Gen Comput (IJNGC) 2(3)
3. Handte M, Becker C, Schiele G (2003) Experiences—extensibility and flexibility in BASE. In: Workshop on system support for ubiquitous computing (UbiSys) at Ubicomp 2003, Seattle, USA
4. Becker C, Handte M, Schiele G, Rothermel K (2004) PCOM—A component system for pervasive computing. In: PERCOM '04—proceedings of the second IEEE international conference on pervasive computing and communications (PerCom'04), IEEE computer society, Washington, DC, USA, p 67
5. Ferscha A, Hechinger M, Mayrhofer R, Oberhauser R (2004) A light-weight component model for peer-to-peer applications. In: 24th international conference on distributed computing systems workshops. Washington, DC, USA, pp 520–527
6. Herrmann K, Rothermel K, Kortuem G, Dulay N (2008) Adaptable pervasive flows—an emerging technology for pervasive adaptation. In: 2008 second IEEE international conference on self-adaptive and self-organizing systems workshops, Washington, DC, USA, pp 108–113
7. Schiele G, Becker C, Rothermel K (2004) Energy-efficient cluster-based service discovery for Ubiquitous Computing. In: EW 11—proceedings of the 11th workshop on ACM SIGOPS European workshop, ACM, New York, NY, USA, p 14. doi:10.1145/1133572.1133604

8. Handte M, Herrmann K, Schiele G, Becker C (2007) Supporting pluggable configuration algorithms in PCOM. In: Proceedings of the workshop on middleware support for pervasive computing (PERWARE), international conference on pervasive computing and communications (PERCOM), New York, USA

9. Object Management Group (2004) The common object request broker: architecture and specification, Revision 3.0.3. online publication, http://www.omg.org/

10. Sun Microsystems (2004) Java remote method invocation specification. online publication, http://java.sun.com/j2se/1.5/pdf/rmi-spec-1.5.0.pdf

11. Haroon M, Handte M, Marron PJ (2009) Generic role assignment: a uniform middleware abstraction for configuration of pervasive systems. In: IEEE international conference on pervasive computing and communications. Washington, DC, USA, pp 1–6

12. Khilwani N, Harding JA, Choudhary AK (2009) Semantic web in manufacturing. Proc Inst Mech Eng B: J Eng Manuf 223(7):905–924. doi:10.1243/09544054JEM1399

13. Obitko M, Vrba P, Mařík V, Miloslav R (2008) Semantics in industrial distributed systems. In: Chung MJ, Misra P (eds) Proceedings of the 17th world congress the international federation of automatic control, Seoul, Korea, vol 17, pp 13880–13887. doi:10.3182/20080706-5-KR-1001.02350

14. Mazzola Paluska J, Pham H, Saif U, Chau G, Terman C, Ward S (2008) Structured decomposition of adaptive applications. Pervasive Mob Comput 4(6):791–806

15. Edwards WK, Newman MW, Sedivy J, Smith T, Izadi S (2002) Challenge: recombinant computing and the speakeasy approach. In: Proceedings of the 8th annual international conference on Mobile computing and networking, ACM, New York, NY, USA, MobiCom '02, pp 279–286. doi:10.1145/570645.570680

16. UPnP Forum (2008) Universal plug and play device architecture, Version 1.0, document revision 24 Apr 2008. online publication, http://www.upnp.org/specs/arch/UPnP-arch-DeviceArchitecture-v1.0-20080424.pdf

17. Sun Microsystems (2006) Jini technology surrogate architecture specification, v1.0. online publication, http://surrogate.dev.java.net/specs.html

18. Karnouskos S, Savio D, Spiess P, Guinard D, Trifa V, Baecker O (2010) Real world service interaction with enterprise systems in dynamic manufacturing environments. In: Benyoucef L, Grabot B (eds) Artificial intelligence techniques for networked manufacturing enterprises Management. Springer, Berlin. ISBN 978-1-84996-118-9

19. Spiess P, Karnouskos S, Guinard D, Savio D, Baecker O, Souza LMSd, Trifa V (2009) SOA-based integration of the internet of things in enterprise services. In: IEEE International Conference on Web Services, ICWS 2009, Los Angeles, CA, USA, pp 968–975. doi:10.1109/ICWS.2009.98

20. Hegering HG, Küpper A, Linnhoff-Popien C, Reiser H (2003) Management challenges of context-aware services in Ubiquitous environments. In: Brunner M, Keller A (eds) Self-managing distributed systems, Lecture Notes in Computer Science, vol 2867. Springer, Berlin, pp 321–339

21. Marin-Perianu M, Meratnia N, Havinga P, de Souza L, Muller J, Spiess P, Haller S, Riedel T, Decker C, Stromberg G (2007) Decentralized enterprise systems: a multiplatform wireless sensor network approach. IEEE Trans Wireless Commun 14(6):57–66. doi:10.1109/MWC.2007.4407228

22. Handte M, Wagner S, Schiele G, Becker C, Marrón PJ (2010) The BASE plug-in architecture—composable communication support for pervasive systems. In: 7th ACM International Conference on Pervasive Services, Newport Beach, CA, USA

23. Aitenbichler E, Kangasharju J, Mühlhäuser M (2007) MundoCore: a light-weight infrastructure for pervasive computing. Pervasive Mob Comput 3(4):332–361

24. Takano M (2004) End-user requirements on industrial networks—issues at end-user site and their potential solutions. In: SICE 2004 annual conference, vol 2, pp 1206–1209

25. Yang M, So SS, Eun S, Kim B, Kim J (2007) Sensos: a sensor node operating system with a device management scheme for sensor nodes. In: Proceedings of the third international conference on information technology: new generations, IEEE computer society, Los Alamitos, CA, USA, pp 134–139 doi:10.1109/ITNG.2007.178

26. Apolinarski W, Handte M, Marron P (2010) A secure context distribution framework for peer-based pervasive systems. In: Pervasive computing and communications workshops (PER-COM Workshops), 2010 8th IEEE international conference on, pp 505–510. doi:10.1109/PERCOMW.2010.5470619
27. Siciliano B, Khatib O (eds) (2008) Springer handbook of robotics. Springer, Berlin

Chapter 2
Deployment and Management of Cooperating Objects

2.1 Overview

Despite the vast amount of past and on-going research in network embedded systems and pervasive computing, real-world deployments of systems of Cooperating Objects are largely still limited to research prototypes. Managing, controlling, and verifying the cooperation and coordination among heterogeneous objects indeed represents a major challenge when the system is deployed in the field. In this respect, Cooperating Objects unfortunately inherit issues germane to some of their constituent technologies, e.g., the lack of visibility into the operation of sensor network systems. This is caused by some of their distinctive characteristics:

- Cooperating Objects are deeply embedded within the real world to perceive and control the environment through sensors and actuators. The dynamics of real-world environments negatively affect the system operation, e.g., because of the unpredictable behaviour of the wireless medium when radio communication is used;
- Cooperating Objects are required to go far beyond the simple interactions found in early deployments of the technologies they build upon. For instance, unlike most sensor network deployments that essentially revolve around pure data collection, cooperating objects are required to form highly dynamic distributed systems with complex interactions;
- Cooperating Objects are often subject to severe resource constraints, e.g., in terms of computation, communication, and available energy. Constrained resources complicate the programming activity, leading to error-prone software, and make it difficult to identify and remedy the causes of such failures.

Contributors of this chapter include: Nils Aschenbruck, Jan Bauer, Armando Walter Colombo, Christoph Fuchs, Philipp Maria Glatz, Stamatis Karnouskos, Paulo Leitão, Marco Mendes, Luca Mottola, Amy L. Murphy, Gian Pietro Picco, and Thiemo Voigt.

S. Karnouskos et al., *Applications and Markets for Cooperating Objects*,
SpringerBriefs in Cooperating Objects,
DOI: 10.1007/978-3-642-45401-1_2, © The Author(s) 2014

The characteristics and issues above entail that the development of systems of Cooperating Objects, their deployment in real-world settings, as well as their management during the system lifetime represent challenging tasks. First steps have been undertaken to address some of these challenges [1], yet these are typically isolated ad-hoc approaches that target only one specific facet of Cooperating Objects. As the current practice is not sustainable in the long-term and already often proved to be insufficient even for small-scale deployments, a widespread adoption of cooperating objects requires a more systematic approach to their deployment and management. Most importantly, not just one specific technology needs to be taken into account, but the focus needs to progressively shift towards the cooperation of heterogeneous platforms.

In this chapter, we discuss how the above issues are being tackled in the deployment and management of real-world systems of Cooperating Objects. We do so by touching upon diverse application scenarios and requirements, cast in a number of real settings:

- Section 2.2 illustrates efforts in employing Cooperating Objects for monitoring railway bridges, pointing out the challenges in data fidelity and distributed processing that need to be overcome for these systems to be practically effective;
- The application of Cooperating Objects in industry automation systems is the subject of Sect. 2.3, where the use of service-oriented architectures is suggested as a way to overcome real-world integration issues;
- Deployment of Cooperating Objects in harsh settings is discussed in Sect. 2.4 for a case of light-weight bird tracking, where weight of the hardware platform and connectivity issues represent the major obstacles to overcome;
- Section 2.5 reports on the use of Cooperating Objects in public safety scenarios, involving diverse computing platforms with distinctly different capabilities and the additional complexity due to mobile settings;
- Finally, Sect. 2.6 illustrates deployments of Cooperating Objects in operational road tunnels, highlighting the challenges stemming from their integration in control systems and with industry strength equipment.

Overall, the rest of the chapter exemplifies the issues at stake in deploying and managing systems of Cooperating Objects. Remarkably, practical and effective solutions in these areas are key requirements for eventual market adoption.

2.2 Monitoring Railway Bridges

2.2.1 Overview

An area where Cooperating Object technology holds great potential is the monitoring of civil structures. A challenging scenario in this domain is given by existing bridges. Particularly, we investigated the application of Cooperating Object technology for monitoring railway bridges in Stockholm, Sweden.

Fig. 2.1 Zolertia Z1 node

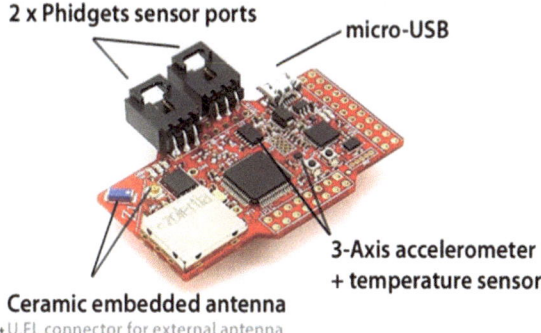

The gradual deterioration and failure of existing structures indeed requires the need for Structural Health Monitoring (SHM) systems to develop a means to monitor the health of structures. Dozens of sensing, processing and monitoring mechanisms have been implemented and widely deployed with wired sensors. On the other hand, the complexity and high installation costs of traditional wired SHM systems have posed the need for replacement with more flexible technology, such as Cooperating Objects.

To counteract memory and energy limitations, thus prolonging the lifetime of battery-operated systems, we designed low-power and memory efficient data processing algorithms. We used in-place radix-2 integer Fast Fourier Transform (FFT). Our implementation increases the memory efficiency by more than 40 % and saves processor power consumption over the traditional floating-point implementations.

A standard-deviation-based peak picking algorithm is next applied to measure the natural frequency of the structure. The algorithms together with Contiki, a lightweight open source operating system for networked embedded systems, are loaded on Z1 Zolertia sensor nodes, shown in Fig. 2.1. Analogue Device's ADXL345 digital accelerometers are used to collect vibration data, to validate the algorithms using supported beam structures.

2.2.2 Application Description/Usage Scenarios

The process of implementing a damage characterization and detection method for engineering structures is referred to as SHM. Although it had been quite a while since the science of SHM was introduced, its use was confined to mechanical structures like airplanes, ships, and machinery. It had never been applied to civil engineering structures until its significance was noticed in the frequent deterioration and collapse of large and prestigious structures. These issues emerge in particular for bridges, whose possible failure may have significant costs, both in economical and social terms.

Fig. 2.2 The catastrophic failure of I-35W Bridge in Minneapolis, Minnesota after collapse on August 1, 2007 (*left*); the Point Pleasant Bridge collapse (*right*)

For example, the catastrophic failure of the I-35W Bridge in Minneapolis, Minnesota (Fig. 2.2 left) and of the Point Pleasant Bridge (Fig. 2.2 right) were among the episodes that alerted the need to devise some means to tell the status of structures before anything could happen. Consequently, a continuous health monitoring of structures is important and a mechanism should be developed by which efficient and accurate information could be obtained.

Researchers, hence, gave special attention to this discipline and proposed their own customized solutions in the last couple of decades, which eventually gave birth to the science of SHM. SHM is thus one of the multidisciplinary fields that integrates the contribution of researchers from mechanical, electrical, civil and architecture engineering. Due to the easy access, the wide availability and reliability of wired systems, many solutions have been implemented using wired sensor networks. However, high installation cost, the need for specially trained professionals for set up and maintenance, and their bulky nature made the research community to divert its attention towards more flexible technologies, of which Cooperating Objects are an example.

Nevertheless, systems of Cooperating Objects deployed for SHM also present significant technical challenges. For example, it is essential to improve the energy efficiency as the energy budget is usually extremely limited, and yet sensed data in SHM applications comes in high volumes that are expensive to transmit wirelessly. On the other hand, memory and computing limitations also reduce the nature and amount of local processing that can be possibly performed aboard the devices to save on wireless transmissions.

Based on the above considerations, we aimed at understanding to what extent existing data processing algorithm can run on Cooperating Object devices in the face of computing and memory limitations, studying the trade-off in terms of quality of the output versus resource consumption. We then designed and implemented customized algorithms to better fit the characteristic constraints of Cooperating Object devices.

Fig. 2.3   Main section of the test beam, with wired monitoring system attached

Fig. 2.4   Wooden support of
the test beam. A Zolertia Z1
node is also visible, attached
at the end of the beam

2.2.3 Key Results and Lessons Learned

Our customized FFT and peak-picking algorithm implementations serve as a foundation to get the study of more complex algorithms started on a sound basis. To that
end, we validated the performance of our implementations against a wired vibration
monitoring system deployed in a university engineering lab.

We used a simple steel beam, shown in Fig. 2.3, supported by wooden blocks at
the two ends. The length of the beam was 3.5 m and the wooden supports, shown in
Fig. 2.4 at the end points add a total of 18 cm. The whole span was further divided into
five sub-parts of each 66.4 cm long. By placing our sensor nodes on these sub-parts,
we collected measurements for real-time and offline analysis.

For the wired system, HBM MGC Plus data acquisition system (DAQ) was used.
The system includes a Si-FlexTM MEMS sensor to be firmly attached to the beam
with a heavy electromagnet attachment, a multichannel ADC and a Catman DAQ
software installed on a laptop computer.

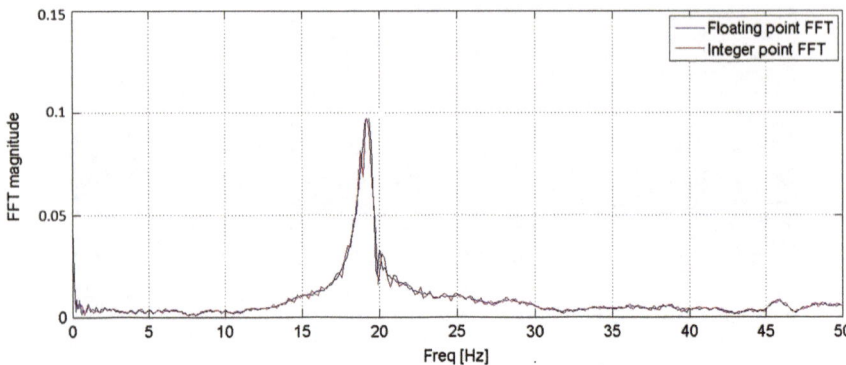

Fig. 2.5 Comparison between Matlab's floating-point FFT and FFT custom implementation for Zolertia Z1

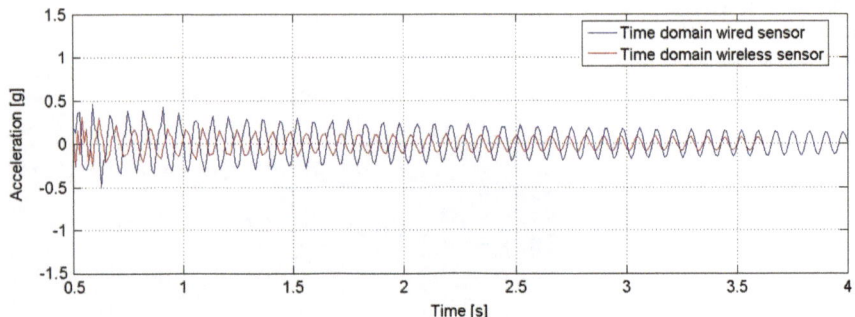

Fig. 2.6 Time domain data from wired and wireless sensors

After collecting time domain data with the wireless sensors, we apply our FFT implementation and compare the result with the floating-point FFT implementation in Matlab. Figure 2.5 shows the two next to each other. Our implementation gives a good approximation of the floating-point FFT on top of saving memory space and reducing energy consumption. Given the low resolution data from the ADXL35 accelerometer, the most important thing is noting the existence of the resonant peak frequencies. These points are what domain experts need to know to extract the important information to feature the behaviour of a bridge.

On the other hand, a time domain plot of the data from the wired and wireless systems is given in Fig. 2.6. As seen from the figure, the time domain data plot from the wired system is relatively of better quality than the one from the wireless system, although the data obtained from the wired and wireless look similar to some extent, which indicates that both are measuring same vibration data from the bridge.

A closer look into the measurements exposes some of the flaws in the wireless sensing system. At low amplitudes, the wireless system introduces noise due to the low resolution (10 bits per sample) of the ADC used in contrast with the 24-bits high resolution one used in the wired system. Here, the 14-bits difference in the fidelity of both data is considered as one of the trade-offs in using wireless sensor system. More specifically, the difference for the signal quality can be attributed to two factors:

- *Noise level*: the Si-Flex accelerometer has lower noise level (300 ngrms/Hz) than that of the ADXL345 accelerometer ($\leq$1.5LSB rms); hence less amount of noise is introduced into the measurements of the wired system.
- *ADC resolution*: the MGC plus DAQ system has a 24-bit ADC which is of much higher resolution than the ADC found in the ADXL345 accelerometer (10 bits of resolution for a g-range of $\pm$2 g); the more the bit resolution of the ADC, the more the quantization level of the converter and eventually, this yields digital samples of higher fidelity.

In summary, we argue that while the software side may already be ready for real-world deployments, on the hardware side we still require better fidelity sensing devices able to get closer to the quality of mainstream wired systems. The provision of such hardware may open market opportunities for this domain, which would instead remain untapped in the current situation.

2.3 Cooperative Industrial Automation Systems

2.3.1 Overview

The future factory is a complex system of systems, where sophisticated and dynamic systems interact with each other in order to achieve the goals at system-wide but also at local level. To realize this, timely monitoring and control as well as open communication and collaboration in a cross-layer fashion are key issues. Modern approaches such as the service-oriented architecture (SOA) paradigm when applied holistically can lead to the desired result [2].

Promising futuristic approaches followed within the EU research projects SOCRADES (http://www.socrades.eu) and IMC-AESOP (http://www.imc-aesop.eu) adopt the "collaborative automation" paradigm where the aim is to develop tools and methods to achieve flexible, reconfigurable, scalable, interoperable network-enabled collaboration between decentralised and distributed embedded systems (Cooperating Objects). In particular, the SOCRADES technical approach [3–6] realized a service-oriented ecosystem, where networked systems are composed by smart embedded devices interacting with both physical and organizational environment, pursuing well-defined system goals. IMC-AESOP empowered by the advanced of cutting edge technologies and concepts [7], pushes the interaction and collaboration capabilities of Cooperating Objects even further by providing an insight how the future industrial automation systems [8] would interact and how their applications would benefit.

In a service-oriented industrial enterprise, the communication and coordination of activities is done by the engineering associated to the service requesting and offer. Cooperating Objects that are integrated in this environment can make usage of the service-oriented mechanism as a mean for cooperation between different autonomous parts of the system. The current application illustrates the deployment and management of service-enabled Cooperating Objects of a pallet-based assembly system (horizontal cooperation), as well as their integration into the production and enterprise resource levels (vertical cooperation). A major benefit of this approach is the modular system deployment and the usage of service-orientation to establish cooperative acts, as well as the integration of heterogeneous resources and information coming from different layers of the enterprise.

2.3.2 *Application Description/Usage Scenarios*

The factory of the future will depend on the services for realizing sophisticated functionalities [2, 5]. Services are basis of a mechanism by which needs and capabilities are brought together, and is a promising way of enabling interoperable interactions among the different cooperating entities. For Cooperating Objects, the principle of service-orientation can be seen as a mean for realizing cooperation. A Cooperating Object represents its actions and resources as a set of services that can be used by other parties e.g., other Cooperating Objects.

As an example, a service-enabled Cooperating Object (as depicted in Figs. 2.7 and 2.8) could be a mediator of a conveyor segment; hence it has the ability to read the sensors and control the actuators of the conveyor, to make it possible to transport pallets from its input to its output. This forms the internal objective of the Cooperating Object, but as it operates in a wider context it has also to respect external/global objectives of the system. The objective and available condition can be offered as a service to the outside (service: transport pallets), so that possible another entity (e.g., a pallet) could request it e.g., *"Please transport me from point A to point B"*. However to complete the service and also to respect global system objectives, the conveyor must interact with the availability service from the next transport unit or workstation connected to its output. This can be seen as the form of collaboration and automatic rearrangement of services in this system.

The approach for creating complex, flexible and reconfigurable production systems is that these systems are composed of modular, reusable entities that expose their production capabilities as a set of services. This composition approach applies to most levels of the factory floor; simple devices compose complex devices or machines, which in turn are composed to build cells or lines of a production system and so on. The same applies to concept of service-oriented production systems and composing complex services from simpler services.

The application scenario that is realized to demonstrate the integration of service-enabled Cooperating Objects is based on a customized Prodatec/FlexLink DAS 30—Dynamic Assembly System. The DAS 30 system is a modular factory

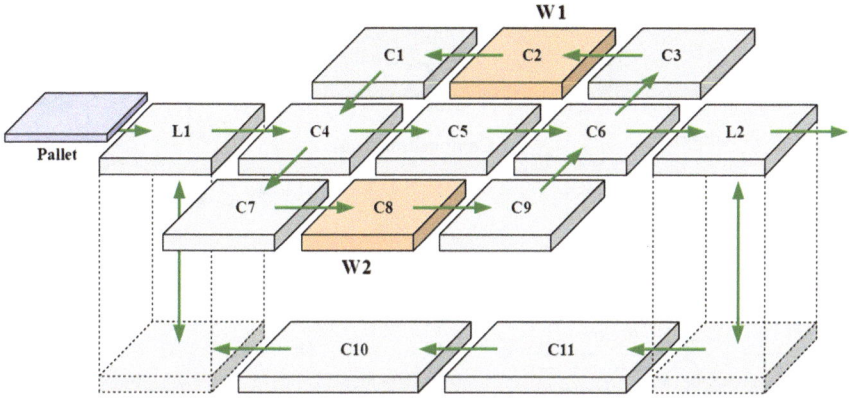

Fig. 2.7 Modular composition of the assembly system

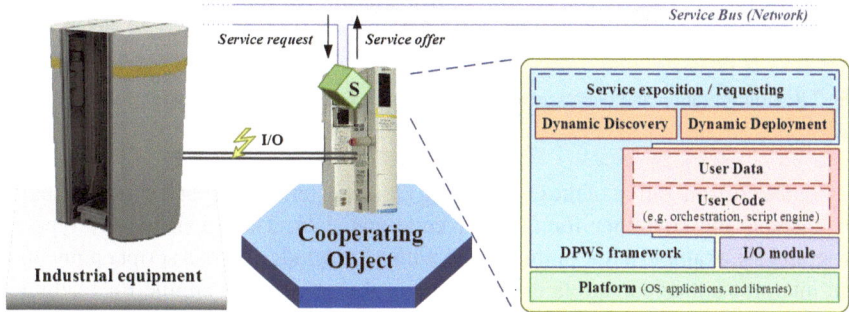

Fig. 2.8 Service-enabled Cooperating Object for industrial automation

concept platform for light assembly, inspection, test, repairing and packing applications. Figure 2.7 shows a representation of the modular composition of the system, using mechanical conveyor modules (C1–C11), lifters (L1 and L2) and workstations (W1 and W2).

The service-enabled Cooperating Objects are the host for most of the services exposed in the system and also responsible for the cooperation and control activities (Fig. 2.8). These devices have two main interfaces: (i) mediating the connection to the shop-floor industrial equipment via I/O (e.g., lifter) and (ii) managing the access to the service bus by exposing and requesting services. The web service infrastructure is based on the SOA4D implementation of DPWS (Device Profile Web Services) (forge.soa4d.org). The Cooperating Object is configurable with the dynamic deployment features available via the SOA4D stack. Once on-line, the Cooperating Object can be discovered (dynamic discovery as defined in the DPWS protocol) and provided services can be requested; similarly it can also request services whenever it needs to.

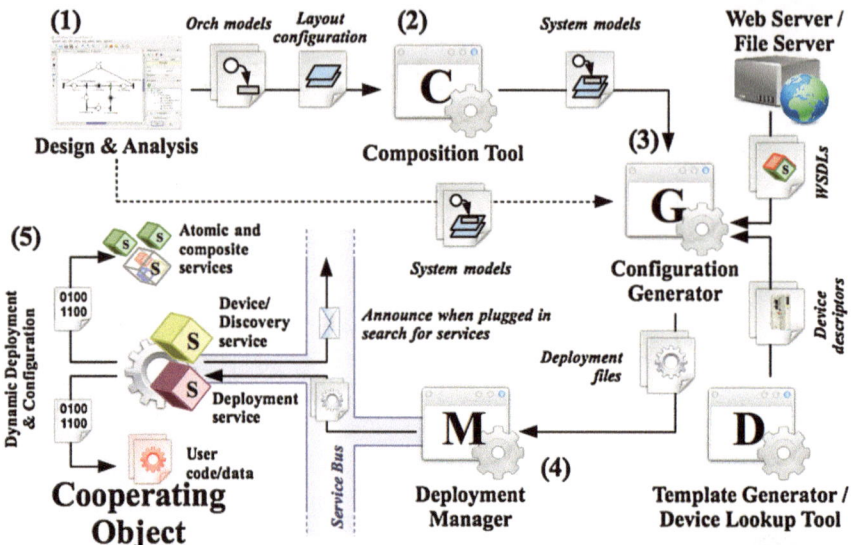

Fig. 2.9 Configuration and deployment tools for service-enabled Cooperating Objects

For the customization of the Cooperating Object in terms of what it should actually do, there is the container for the user code and data. These can be any type of program, for example a reasoning engine with a knowledge base, a script engine that can interpret uploaded scripts or a model-based orchestration engine for reading a given work-plan made of services (an orchestration) and execute it. The last option is used in this work as reference for the system deployment and management. The tool chain needed for composing systems, creating configuration files and deploying those files to Cooperating Objects is depicted in Fig. 2.9, which illustrates the complete engineering sequence from the design of the components or the system, passing by the composition and followed by the deployment to the devices.

As depicted in Fig. 2.9, the sequence starts with the design of component work-flow/model (*step 1*) once per device type. Composition of the instance models can be done for one or more system model(s) (*step 2*) followed by the generation of configuration files (*step 3*). This process generates basically two files: (i) a service class descriptor, containing the referenced port types and a model representation, and (ii) and a device descriptor, containing the device and hosted service information, including all discovery hints needed by the execution engine (later on to resolve the referenced component services). The Cooperating Object must be running and ready to receive the configuration (*step 4*). Then the deployment manager id invoked in order to download the descriptor files for a specific system model to the target device. This step is repeated until all models are deployed. A new device is generated if there is server information in the deployment files. The new device can then be used by any client. Once a target has received the configuration and configured correctly, the execution start automatically (*step 5*) by first making announcement and discover

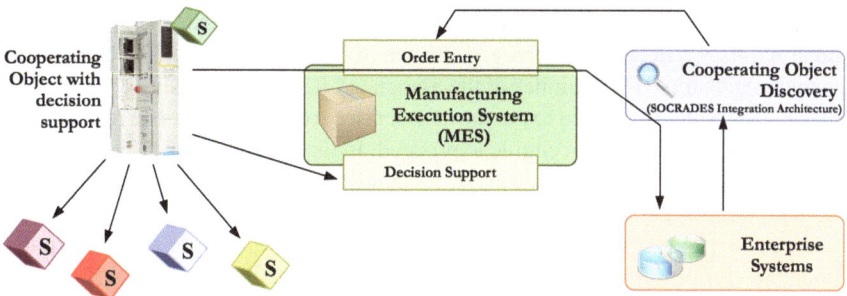

Fig. 2.10 Manufacturing execution system overview

of needed services (dynamic discovery) and proceeding afterwards to the operation defined by the orchestration engine.

As can be seen, the current approach is production agnostic, since it does not contain any information concerning products. This is intended for the separation of concerns, i.e., production and resource management is handled by other Cooperating Objects or services in the system. For production management and information in adjacent to the automation tasks, production orders are integrated via service-orientation from the enterprise resource planning (ERP) system directly on the shop floor. The detailed production steps are stored in the Manufacturing Execution System (MES). This MES interacts with the Cooperating Objects via an orchestration engine and also is in contact with the ERP system as shown in Fig. 2.10.

As mentioned, the depicted system utilizes collaboration among multiple layers i.e., from the devices in the shop-floor, to the MES and the ERP. However to make this possible two issues need to be resolved i.e., (i) dynamic discovery of Cooperating Objects and their services that do not reside on the same physical location or network segment, and (ii) seamless interaction among these Cooperating Objects. To resolve this, we have developed a middleware named SIA (SOCRADES Integration Architecture) [9] developed explicitly with "device-to-business" integration in mind [5]. The middleware offers auxiliary services in interacting with devices and systems. As a complementary functionality a network application (named LDU) is created, that provides discovery of devices and services via DPWS (Device Profile for Web Services) on the local network and connection to the enterprise system. The application is cross-platform ready (prototype is implemented in Java) and hence can be automatically instantiated by a Cooperating Objector even run manually in the local network by the factory operator (by just clicking on it in his web browser). Different versions of the application can add-up functionalities, e.g., proxy also specific enterprise services at the local shop-floor, where they can be discovered and used by the Cooperating Objects as well as other devices and services. The middleware provide a means for connecting and managing devices from different physical and network premises.

The system operates autonomously with information shared among the Cooperating Objects which may spawn various levels e.g., local sensors, MES, enterprise systems and services. For instance once the workflow is started, the orchestration engine requires a decision to proceed further. To identify the pallet to be handled, the engine gets the RFID number of the associated RFID reader using the matching service. This pallet ID is used to get the next service from the MES. The returned service allows the orchestration engine to proceed. In turn, the pallet is moved on to the determined facility, e.g., workstation 2. Reaching the destination, the accompanied production unit is called to execute a service for the given pallet ID. It is possible to let the system produce the units automatically and allow completing an order without human interaction. This scenario focusing on cooperation happening horizontally at "device" level as well as vertically i.e., among devices and systems/services is part of ongoing efforts to show the added-value benefit that can be materialized with new disruptive technologies and concepts in the domain of industrial automation.

2.3.3 Key Results and Lessons Learned

Taking the granularity of intelligence to the device level allows intelligent system behaviour to be obtained by automatically composing configurations of devices that introduce incremental fractions of the required intelligence. This approach favours adaptability and rapid reconfigurability, as re-programming of large monolithic systems is replaced by reconfiguring loosely coupled embedded units, that can then further enhance their functionalities via cooperation with other devices, systems and services.

The realized service-oriented solution demonstrated the viability to develop complex distributed and Cooperating Object enabled applications, based on service-orientation mechanisms that allow the horizontal and vertical integration. In fact, the service-oriented principles allow to overcome interoperability problems that usually appear in these environments due to the existence of heterogeneous software and hardware applications. A key result demonstrated is related to the flexibility and modularity exhibited by the service-oriented based approach to develop Cooperating Objects solutions. In fact, the on the fly adaptation to unexpected disturbances or even to process changes (in these cases requiring off-line adaptation) is a crucial factor for the success of this kind of solutions.

From a functional perspective, the focus is on managing the vastly increased number of intelligent devices and mastering the associated complexity. From a run-time infrastructure viewpoint, the focus is on a new breed of very flexible cooperative real-time networked embedded devices (wired/wireless) that are fault-tolerant, reconfigurable, safe and secure. Especially auto-configuration management is a new challenge that is addressed through basic plug-and-play and plug-and-run mechanisms.

From technological and infrastructural viewpoints, the use of the Service-Oriented Architecture (SOA) paradigm implemented through web service technologies enables the adoption of a unifying technology for all levels of the enterprise, from sensors and actuators to enterprise business processes. This means that low cost devices may communicate directly to higher-level systems and enhance their own functionality, which may lead to more adaptive and lightweight approaches.

As already probably noticed, only minimal assumptions about the concrete production line are present in the whole system design. It is important to recall here that the operational behaviour of the devices is self-controlled and guided by internal/external events that also may correspond to service calls. Collaboration among the stakeholders is made possible with dynamic discovery and seamless interaction among them. Additionally more sophisticated approaches can be realized based on orchestration of the existing Cooperating Objects in other Cooperating Objects as well as with the support of the infrastructure [5]. We have also to point out that this approach is not limited to the specific depicted example case, but other domains and systems may also benefit from it, leading to a continuously evolvable infrastructure [2] that may adapt to the business needs and hence provide a competitive advantage.

2.4 Light-Weight Bird Tracking Sensor Nodes

2.4.1 Overview

There are mainly two methods in use for tracking the behaviour of birds. Birds are either tracked by means of applying a sensing unit or some sort of marker onto individual birds or they are manually tracked by people spotting flocks of birds or maybe individuals of rare species. Unfortunately, both ways are tedious ways for approaching the problem, in particular, due to the manual labour involved for recollecting units and the uncertainty in observations when it comes to gathering continuous traces. Also, these methods are not always sufficient from an application point of view. For instance, it is difficult to track the influence of man-made structures on the behaviour of birds that way, which is often necessary information for biologists studying bird species. The UvA Bird Tracking System (BiTS) aims at overcoming such issues. A reconfigurable bird-tracking platform has been deployed that allows long-term application of harvesting-enhanced GPS-enabled sensor nodes with wireless readout. The possibility to download birds' data through permanently installed base stations feeding into a database with a 'virtual lab' web front-end allows collecting long-term high quality traces of birds.

Though the system is readily available and working, implications stemming from the challenging application context make up a number of interesting research questions that need to be addressed. Designing a structure that is to be deployed on birds puts severe constraints on weight, size and durability which in turn translates into limited performance (i.e., limited data that can be collected or that can be downloaded by a base station). In a joint effort with UvA, The research aim is targeting research and evaluation of cooperative networking behaviour among sensor nodes on different birds.

Fig. 2.11 UvA BiTS mote with programming interface exposed and a BlueBean development kit for radio evaluation networking are shown

Intelligent data dissemination needs to be applied in the context of intermittently connected networks to improve system performance. This translates into the need for accurately profiling and modelling the system as well as setting up realistic and highly scalable simulation environments that need to be tested with hands-on experience from real-world experiments. This way, applicability of novel protocols can be tested based upon existing traces and simulated behaviour can be checked by means of comparison of its characteristics with results from real-word experiments. Figure 2.11 shows an actual hardware prototype and part of a development kit for evaluating networking protocols. Outcomes from using a simulation environment will be discussed later.

2.4.2 Application Description/Usage Scenarios

As shown in Fig. 2.12, the Uva BiTS architecture comprises multiple bird tracking applications and combines their information in a database. Readily completed applications include a case study on Common Buzzard flight activities at an airport, migration behaviour of seagulls and their interaction with wind farms. Exemplary results from tracing flight activities at an airport are shown in Fig. 2.13.

For bird tracking applications at airports, options like manual bird spotting, using marking or sensing devices without radio communication are out of the question due to hard real-time constraints. Designing a bird monitor, tracking e.g., flocks of geese flying low and close to an airport, requires real-time information for avoiding potentially dangerous bird strikes. Employing a monitor by means of the UvA BiTS platform might allow to not necessarily track flocks of birds but also to possibly recognize encounters with base stations deployed around the airport.

One of the main application symbolized in Fig. 2.12 is monitoring foraging and tracking migration of birds. Several applications have been implemented for foraging (e.g., tracking Oystercatcher in the Dutch Wadden Sea) and for studying cross-continental migration behaviour. For tracking foraging behaviour, it is particularly

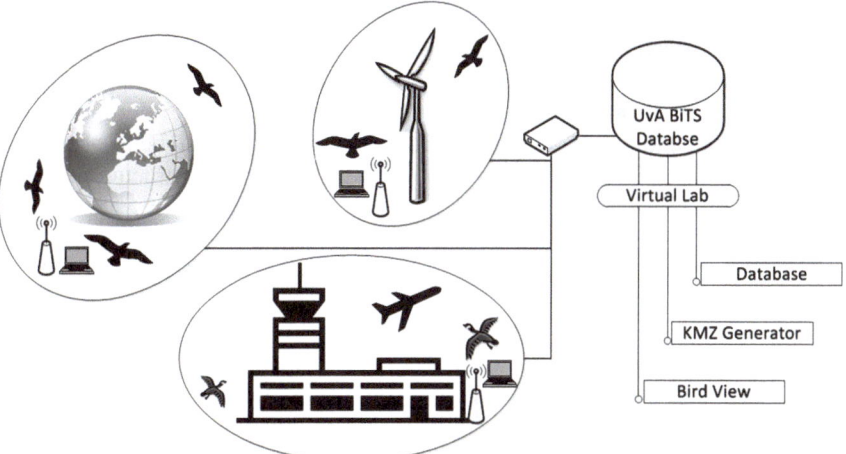

Fig. 2.12 Multiple applications are run side by side eventually sharing base stations

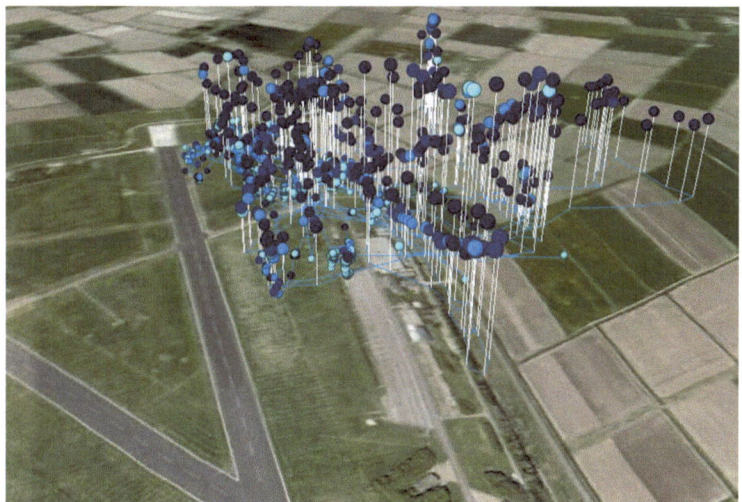

Fig. 2.13 GPS trace of Common Buzzard flying at an airport. Taken from [10]

beneficial to have a base station close to where motes have been deployed. For the case of tracking migration behaviour it usually gets more complicated. Memory load gets higher between consecutive readouts and cooperative readout among multiple base stations can be beneficial.

Last but not least, the interaction of birds with man-made structures is being monitored. Successful traces of seabird-windfarm interactions have been taken at Orford Ness close to the east coast of England. 25 Lesser Black-backed Gulls have been attached nodes of the UvA BiTS platform in 2010 and 2011. Interaction with wind farms can be an important issue to be considered when it comes to conservation of birds' natural habitats. Plans had been made to set up windfarms at Brenner in western Austria being one of the main transit routes through the Alps. However, it also one of the main migration routes for birds crossing central Europe through the Alps. Information that could be collected with the BiTS could be invaluable for studying how bird migration routes are impacted from men-made structures, but also from long-term changes of climate conditions.

2.4.3 Key Results and Lessons Learned

System development is challenging because of issues with long-term application life cycles compared to short-lived state-of-the-art in the rather new field of harvesting-enabled sensors with hard-to-predict mobility. Setting up a bird-tracking application may take years of planning and testing of a deployment, because of yet-to-be validated yearly migration patterns of birds. However, history tells us that hardware components and protocols can be utilized in ever more efficient ways—a couple of years ago UvA BiTS might not have been considered feasible—which does not match the speed of possibly adapting the application at hand. Running the system for years, the information that is traced by birds becomes more and more valuable (i.e., long lasting traces allow for more expressive studies of changing bird behaviour) and therefore newly deployed platforms/applications must not conflict (i.e., be backwards compatible) with older applications that are still running. This mismatch between innovation in technology (of course the costumer wants to work with the highest possible performance for any new platform) and long application lifecycle demands for careful consideration at every new design step. Existing compatibility between applications and versions of platforms must not be violated and future compatibility must not be hindered either—a major demand in the domain of Cooperating Objects.

As being a key element for any wireless Cooperating Objects, communication standards and respective hardware components and protocol implementations deserve special consideration when it comes to compatibility in long-term deployments. In particular in the case of hard-to predict mobility, collecting and reprogramming sensors is not applicable means of avoiding having outdated instances of the platform in the field. Due to those issues being described above, SerialNet (providing versatile reconfiguration features) over ZigBee-compliant (with ZigBee being the only de-facto small-footprint wireless industry standard widely in use) ZigBit running on a far-spread AVR-based platform (enforcing long lasting support and further development of the system).

Currently, wireless communication capabilities are made use of for wireless read-out functionality and eventual multi-hop communication for the non-intermittently-connected case. This means that first of all, direct (single hop) readout is performed, and furthermore, readout via relaying nodes (ZigBee router functionality) is possible as well as long as the end-to-end channel is connected and does not get blocked by other communication. This raises the question whether data good put could be

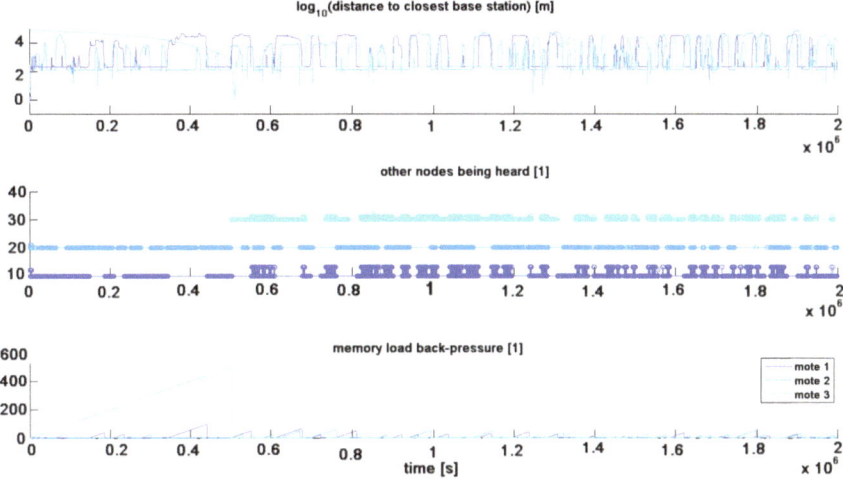

Fig. 2.14 Sample evaluation of three gulls' traces around a base station in Texel (The Netherlands)

increased by means of data muling or message ferrying as known from other sensor network applications encountering the intermittently connected (delay tolerant) case. Due to the dependable nature of the BiTS application, it is worth it to first study birds' mobility and interaction characteristics before deploying a novel networking approach that is possibly degrading existing deployments' performance. Figure 2.14 depicts a snapshot of a possible scenario rerunning traces that have actually been gathered in the wild and three motes' connectivity among each other and to a base station while simulating (using a modified Castalia) a wireless readout protocol similar to the one being used in the actual application.

Figure 2.14 shows the motes' distance to their closest base station in the uppermost plot. Next, other nodes or base stations that could be recognized are being indicated. Circles at 10/20/30 indicated base station contact, where circles at 10/20/30+i indicate contact to node 'i'. The lowermost plot compares the memory load of different motes given that GPS readings are taken at a constant rate and the mote with the higher memory load is allowed to offload data if multiple nodes are in base-station range at the same time. Three key observations can be made from looking at this snapshot of normalized information.

- There would possibly be ways of improving good put performance if the birds' movements would have been known in advance. Thus using delay tolerant networking approaches might be beneficial for the application. However, irregularity in the application scenario may drastically change the constraints which need most attention at a time, as can be seen with the peak load of memory at mote 3—not to mention the variability of available energy which has been omitted from the plots.
- However, different bird (and even different researcher) behaviour may vary and lead to different performance among different runs with one and the same protocol.

Obviously, mote 3 was activated at a way different place than the other motes, though belonging to the same deployment. Furthermore, mote 2 has hardly any contact to other birds' motes, though they all have contact to the same base station on a regular basis.

• The main lesson being learned when inspecting real-world traces is that one is dealing with a highly dependable system, where it is difficult to tell in advance, what constraints will be put to their limits to what extent and how often. Therefore, extensive simulation is necessary before protocol changes or architectural modifications can be carefully incorporated into the system.

Despite the fact that numerous problems have successfully been overcome as results of many deployments have shown, there are still a couple of optimizations that are currently being researched. In particular energy and memory efficiency and capacity receive attention as does networking and data communication good put.

2.5 Public Safety Scenarios

2.5.1 Overview

Public safety organizations such as first responders, fire fighters, police, and military units need robust communication networks to cooperate and transmit different kind of sensor information. These networks have to be reliable even when a pre-deployed infrastructure has been destroyed. Wireless multi-hop networks (such as Mobile Ad-Hoc Networks (MANETs), Wireless Sensor Networks (WSNs), and Wireless Mesh Networks (WMNs)) promise to meet the requirements of (1) spontaneous deployment, (2) being independent of any kind of existing infrastructure, and (3) robustness in the sense of self-organization and self-healing by their very definition. These networks have been a topic in research for more than a decade now. Recently, real-world tests and deployments provided valuable insights concerning challenges and future research directions. There are different mesh and WSN testbeds (e.g., [11–13]) enabling the research community to run tests in static and mobile real-world networks. However, concerning Cooperating Objects in public safety scenarios, there are significantly different requirements:

• Spontaneous deployment,
• Mobility typical for public safety scenarios,
• Typical applications and traffic for public safety scenarios.

Due to these characteristics, deploying Cooperating Objects in public safety networks is a huge challenge. To overcome this challenge, we have developed *Bonn Sens* [14, 15] a prototype based on commercial off-the-shelf (COTS) hardware. The prototype comprises typical public safety applications and is spontaneously deployable. Furthermore, this prototype enables us to perform evaluations with real public safety end-users, e.g., by deploying the prototype in manoeuvres.

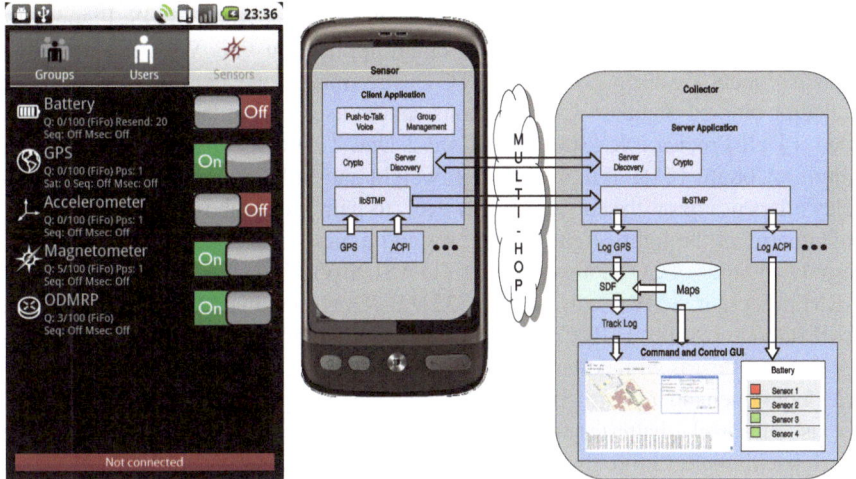

Fig. 2.15   BonnSens System

2.5.2 Application Description/Usage Scenarios

In public safety scenarios there are two common requirements for a command and control system: (1) push-to-talk voice communication and (2) map-based (blue-force) tracking. To communicate inside a team, squad, group, and platoon as well as in between, talk-group based voice communication is important. In addition, the central as well as local command and control points need to know where their units are.

A suitable Cooperating Objects architecture for this scenario consists of a distributed sensor and collector application for the transmission of sensor data over a wireless multi-hop network as shown in Fig. 2.15. The Cooperating Objects support the collection of sensor information (e.g., Global Positioning System (GPS), accelerometers, and magnetometer) via modular extensible plugins. The transmitted sensor data is stored in a database by the collector on the central side. Depending on the type of sensor, the data can be visualized using different types of Graphical User Interface (GUI). Some kind of data, such as positioning data, may be additionally processed by a sensor data fusion algorithm before being visualized on a map. The voice communication is realized using a peer-to-peer based voice application including a dynamic group management on the lightweight nodes.

The architecture consists of two components: (i) portable, lightweight sensing objects and (ii) fully-equipped collector objects. For the sensor objects standard COTS smartphones are used. Smartphones available on the market today, often come with integrated sensors like GPS, accelerometer, and compass. These devices are an ideal basis for the lightweight objects. However, smaller WSN mote like objects can in principle be used as well.

We have implemented a client sensing application for Android OS. Figure 2.15 shows a screenshot of our sensor data client application *BonnSens*. For the light-weight objects, we currently use the HTC Desire smartphone. For the fully-equipped collector objects, standard COTS laptops can be used. In the last deployments, we used a Dell Precision M4300 (Intel Core 2 Duo T7700 2,4 GHz and 4 GB RAM) running Ubuntu 10.04. All objects (lightweight and fully-equipped) can be used as relays depending on the topology. However, to gain more robust topologies and to safe energy at the lightweight objects, we add a mesh backbone. For the mesh, we tested two kinds of COTS mesh routers: (1) ASUS WL-500g Premium V1 (266 MHz ARM processor, 8 MB Flash memory, 32 MB RAM) and (2) ALIX 3D2 (500 MHz AMD Geode LX800, 1 GB Compact Flash memory, 256 MB RAM). In both routers we use WiFi-cards with Atheros chipsets (TP-LINK TL-WN660G). For an easy on-site deployment an infrastructure-independent power support is a requirement. Thus, we use motorbike-batteries with 12 V-20 Ah. Using these batteries, we can run the mesh backbone for more than 12 h without any infrastructure.

For the relaying we implemented a multicast routing approach, as both core appli-cations, voice communication and blue-force tracking, imply multicast. The voice application is group-based. Thus, it can be efficiently realized by multicast-groups. In some public safety scenarios, there may be several fully-equipped objects with a demand for a visualization. As multicast routing protocol we chose the reactive On-Demand Multicast Routing Protocol (ODMRP) [16]. ODMRP is a mesh-based approach based on scoped flooding. A selected subset of nodes forwards the packets. We chose ODMRP as it showed promising results in public safety specific simula-tive performance evaluations [17]. Furthermore, it showed to behave quite reliable even under attacks like sinkholes, as the mesh structure provides robustness against the attraction of routes. We implemented ODMRP using the Click modular routing framework [18]. The Click user-space mode enables us to run ODMRP on the mesh routers as well as on Android phones.

To provide functionalities for sensors such as registering at a receiver, timestamp-ing of sensor data, synchronization of all nodes, as well as providing sensor manage-ment functions, we specified and implemented the Sensor Data Transmission and Management Protocol (STMP) [19]. In order to realize a consistent implementa-tion of STMP for lightweight as well as for fully-equipped objects, we modularized STMP by implementing it as a commonly usable library in C named *libSTMP*. Client applications may thus be programmed either using the device specific API (e.g., Android API for the deployment on smartphones) or using native C code (e.g., for the deployment on a laptop) both accessing the same STMP library.

To support tracking of a sufficient accuracy even in complex scenarios such as urban canyons, appropriate sensor data fusion algorithms have been integrated. As shown in [20] a standard Kalman filter [21] may suffer significantly from Out-of-Sequence (OoS) measurements. An Accumulated State Density (ASD) methodology [22] which allows to calculate the impact on all states within a given time window proved to be a valuable alternative.

2.5.3 Key Results and Lessons Learned

We present the lessons learned during our deployments. To get feedback on our approaches as early as possible, we started to deploy early versions of our architecture in manoeuvres. We deployed mesh backbones as well as portable, lightweight sensing nodes. The latter ones are smartphones carried by the units. To date, we have done spontaneous deployments in four manoeuvres in cooperation with the Johanniter Academy in Münster (Germany) on the manoeuvre ground of the institute of fire-fighting. During these manoeuvres, we had to learn several lessons that we will discuss now. By doing so, we will also describe challenges typical for public safety networks and systems.

As portable, lightweight sensing objects, smartphones are used in our system. Many smartphones on the market today have integrated sensors like GPS. Thus, these devices seem to be an ideal basis for the map based tracking part of our command and control system. However, just visualizing raw GPS positions of all lightweight objects is not sufficient for the requirements in public safety scenarios. An accuracy of 1m is typically requested. Such an accuracy is challenging on typical manoeuvre sites due to obstacles, etc. that may temporarily prevent the proper reception of a GPS signal. Thus, the raw positions have to be filtered and fused. Using standard filters such as a Kalman filter in the system yields to new challenges such as OoS measurements (cf. [20]). The filters need to be optimized for the usage in typical multi-hop networks.

For the measurements as well as for the units that want to use the command and control system, it is important that all objects do not run out of battery. If necessary, single batteries have to be exchanged. For doing so, it is important to know the objects that have battery problems. Thus, we learned that the most important sensor data to be aware of during a deployment is the battery power of the devices deployed. Furthermore, energy awareness in general is very important.

Some voice messages and sensor information transmitted may have higher importance than others. This becomes relevant especially when the data rate is limited due to sub-optimal signal propagation characteristics or resource constrained nodes. Furthermore, when messages are transmitted as broadcasts on the link layer, the basic rate is used. This may lead to additional rate limitation. Thus, data prioritization or, more general, communication and sensor management needs to be implemented. In our system, we implemented the Sensor Data Transmission and Management Protocol (STMP) [19] to take care of prioritization as well as communication and sensor management.

Especially in early deployments, it is important to save all data locally as well—just in case there is a problem with the network (e.g., limited data rates). This also allows for an easier debugging. However, when relying on local logs, non-synchronized clocks may be a challenge. Furthermore, approaches well-described in the literature and evaluated in simulations or labs, may not run very well in real deployments. For example, links may be extremely variable which yields suboptimal performance of some approaches.

2.6 Road Tunnel Monitoring and Control

2.6.1 Overview

State-of-the-art solutions for road tunnel lighting either use pre-set light levels based on date and time, or adjust the lights based on an open-loop regulator relying on an external sensor. Both solutions disregard the actual lighting conditions inside the tunnel, and may endanger drivers or consume more power than needed. The solution developed within the TRITon (*T*rentino *R*esearch and *I*nnovation for *T*unnel M*on*itoring, triton.disi.unitn.it) project deployed a WSN along the tunnel walls to measure the light intensity and report it to a controller, which closes the loop by setting the lamps to match the lighting levels mandated by law. Unlike conventional solutions, our system adapts to fine-grained light variations, both in space and time, and dynamically and optimally maintains the legislated light levels. This enables energy savings at the tunnel extremities, where sunlight enters, but it is also useful inside the tunnel to ensure the target light levels even when lamps burn out or are obscured by dirt. The overall architecture for adaptive lighting, of which the WSN is an integral element, was awarded a European patent in March 2012.

The system was developed with the goal of reducing the management costs of road tunnels and improving their safety. Our WSN-based control system has been installed since August 2010 in an *operational* tunnel on a high-traffic freeway, where it has been running without intervention. Our contributions range from hardware to software, with the former built on top of our TeenyLIME middleware. Based on measurements and calculations, the energy consumption in our tunnel is up to 50 % less than a solution with standard technologies.

2.6.2 Application Description/Usage Scenarios

The system as shown in Fig. 2.16 contains many components working in concert to monitor and control the loop. The principal element collecting the light values inside the tunnel is a WSN composed of approximately 90 nodes divided between the two carriageways of the tunnel. The sensed values are collected by four gateways, combined with the value of an external luminance sensor and sent to an industrial PLC. The PLC implements a centralized control logic to determine the lamp levels required to meet the legislated levels and sends commands to the individual lights to establish these levels. In addition, a SCADA subsystem is connected to the PLC and provides an interface to the human operator for visualization and manual control.

The WSN nodes are functionally equivalent to TelosB motes, equipped with an MSP430 microcontroller, a Chipcon 2420 radio chip, and an on-board inverted-F microstrip antenna. A custom sensor board is attached and contains 4 ISL29004 digital light (illuminance) sensors, and 1 TC1047A temperature sensor. Each node

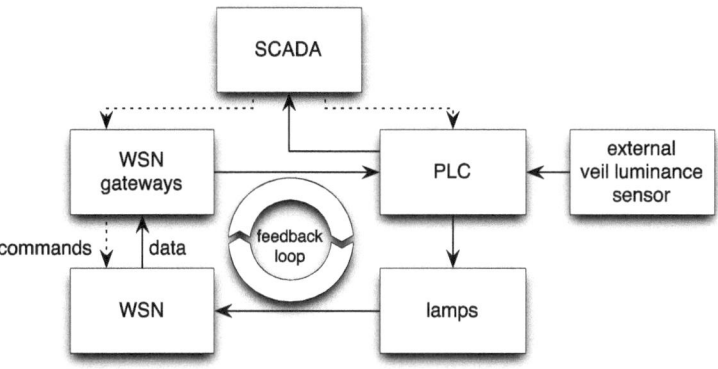

Fig. 2.16 Tunnel System Architecture

is powered by 4 Duracell Procell D-size batteries and placed inside an IP65 (water and fireproof) polycarbonate box with a transparent cover.

We also placed four Verdex-Pro embedded computers by Gumstix to serve as the gateway nodes collecting light samples. These are powered and connected via Ethernet to the PLC and SCADA.

The software installed on the motes runs on top of TinyOS [23]. However, unlike the vast majority of other deployments where application and system software sit directly on the operating system, we built on top of the TeenyLIME [24] middleware (teenylime.sourceforge.net).

This choice was motivated by the fact that TeenyLIME was used successfully by our group in another long-term, real-world deployment for structural health monitoring [25], where its higher-level abstractions were shown to reduce the overall code footprint, allowing one to pack more functionality on memory-restricted nodes.

The main software components are for sensing, data collection and data dissemination. Sensed light values are reported by each node every 30 s, however the reported value is computed over multiple samples from each of the four sensors, eliminating outliers and averaging the remaining samples. To collect the data over multiple hops, we implemented our own tree-based collection protocol that uses LQI to measure link quality. This is motivated by the experimental observation that the resulting trees are similar to those obtained with ETX-based protocols, but with much less overhead [26]. Our protocol also supports multiple sinks, with the best sink being identified implicitly by choosing as parent the neighbour with the smallest node-to-sink routing cost. By periodically refreshing the tree, the routing paths adjust to changes in the topology and all nodes will automatically recover in case a sink fails. Finally, we use a hop-by-hop recovery mechanism to ensure data reliability. The last component, data dissemination, is used to dynamically reconfigure system parameters such as the sampling frequency.

2.6.3 Key Results and Lessons Learned

Our experiences from more than two years of working in actual tunnels are reported next.

Our final deployment is on a high-traffic road, therefore carrying out experiments on this site would have been impractical due to the need to partially or totally block the road. Therefore, we obtained access to a shorter, lower-traffic tunnel that served as a testbed we could more easily access to run experiments. These tests were critical to test the stability of the system, evaluate the energy consumption, and establish critical application-level parameters. In this test scenario we were able to identify critical problems in both the hardware and software which could not have been detected in a testbed environment. For example, the mounting mechanism attaching the original sensor board to the mote was affected by vibrations caused by traffic in the tunnel. The sensor boards installed in the final system have a different mounting mechanism that does not suffer from this problem. While the tests were invaluable, there were still differences between the test tunnel and the final tunnel. Specifically, in the test tunnel, 44 nodes were spread along 132 m instead of the target 630 m. While we performed a few tests with the target node density to understand the behaviour of the routing protocol, we could not anticipate all issues that arose in the final test tunnel. For example, while our estimates for node lifetime were accurate, we saw an unexpected pattern of node death in the final tunnel.

We designed our system to support the needs of the adaptive lighting application. However, in the course of the project, we used the same system to test an innovative system for fire detection as well as to monitor carbon monoxide levels inside the tunnel. The former was done with minor modifications to the original system, e.g., increasing the sampling frequency and separating out the infrared component from the light sensors. By doing so, we were able to accurately track the location of a propane flame on the back of a firetruck as it moved through the tunnel. For CO detection, we designed a new sensor board with a CO sensor, then used the backbone of light sensors to transmit the CO data to the gateways.

As this system is installed in a real road tunnel, controlling lights that fundamentally affect the safety of the drivers, it is important that the system function at all times. While it is possible to signal the failure of a single node, indicating that it should be repaired, the system must continue to function. In case of significant failures, the system can fallback to using whole-tunnel pre-set light levels, but single node failures must be compensated. Therefore, the node density must be sufficient that a single node failure does not disconnect the tree. Further, we introduced multiple gateways in each tunnel in order to survive single gateway failures.

As designed, our system has node lifetimes above one year. However, additional studies done by our group have shown that we can increase lifetime by applying model-driven data acquisition, a technique that aims to reduce the amount of data reported by each node. At each node, a model predicts the sampled data; when the latter deviate from the current model, a new model is generated and sent to the data sink. With a simple derivative-based prediction model that is easy to implement on the limited capacity WSN nodes, our experiments suggest that the system lifetime can be tripled [27].

2.7 Conclusions

As a result of the efforts described above and of the general state of the art, many of the research questions related to deployment and management of Cooperating Objects are now well understood, while the corresponding solutions are still slowly making it into mainstream practice.

The missing tile to the puzzle, which is going to boost the acceptance of Cooperating Objects technology and correspondingly open new market avenues, lies in the standardization and integration of the methodologies at stake, which still partly represent ad-hoc or isolated efforts. Establishing a sound basis in the deployment and management of Cooperating Objects will indeed lessen the burden to create products out of research prototypes.

References

1. Mottola L, Picco GP (2011) Programming wireless sensor networks: fundamental concepts and state of the art. ACM Comput Surv 43(3):19:1–19:51. doi:10.1145/1922649.1922656
2. Colombo AW, Karnouskos S, Mendes JM (2010) Factory of the future: a service-oriented system of modular, dynamic reconfigurable and collaborative systems. In: Benyoucef L, Grabot B (eds) Artificial intelligence techniques for networked manufacturing enterprises management. Springer, New York. ISBN 978-1-84996-118-9
3. Karnouskos S, Baecker O, de Souza LMS, Spiess P (2007) Integration of SOA-ready networked embedded devices in enterprise systems via a cross-layered web service infrastructure. In: Proceedings of IEEE conference on ETFA emerging technologies and factory automation, pp 293–300. doi:10.1109/EFTA.2007.4416781
4. Colombo AW, Karnouskos S (2009) Towards the factory of the future: a service-oriented cross-layer infrastructure. In: ICT shaping the world: a scientific view, ISBN: 9780470741306, European Telecommunications Standards Institute (ETSI), Wiley, Sophia-Antipolis
5. Karnouskos S, Savio D, Spiess P, Guinard D, Trifa V, Baecker O (2010) Real world service interaction with enterprise systems in dynamic manufacturing environments. In: Benyoucef L, Grabot B (eds) Artificial intelligence techniques for networked manufacturing enterprises management, ISBN 978-1-84996-118-9. Springer, London
6. Taisch M, Colombo AW, Karnouskos S, Cannata A (2009) Socrades roadmap: the future of soa-based factory automation. http://www.socrades.eu/Documents/objects/file1274836528.84
7. Karnouskos S, Colombo AW (2011) Architecting the next generation of service-based SCADA/DCS system of systems. In: 37th annual conference of the IEEE industrial electronics society (IECON 2011), Melbourne
8. Karnouskos S, Colombo AW, Bangemann T, Manninen K, Camp R, Tilly M, Stluka P, Jammes F, Delsing J, Eliasson J (2012) A SOA-based architecture for empowering future collaborative cloud-based industrial automation. In: 38th annual conference on the IEEE industrial electronics society (IECON 2012), Montréal, Canada
9. Spiess P, Karnouskos S, Guinard D, Savio D, Baecker O, Souza LMSd, Trifa V (2009) SOA-based integration of the internet of things in enterprise services. In: Proceedings of IEEE international conference on web services, (ICWS 2009), Los Angeles, pp 968–975. doi:10.1109/ICWS.2009.98

10. University of Amsterdam (2012) UvA bird tracking system. http://www.uva-bits.nl/other-projects/
11. Bicket J, Aguayo D, Biswas S, Morris R (2005) Architecture and evaluation of an unplanned 802.11b mesh network. In: Proceedings of the 11th ACM internernational conference on mobile computing and networking
12. Raychaudhuri D, Seskar I, Ott M, Ganu S, Ramachandran K, Kremo H, Siracusa R, Liu H, Singh M (2005) Overview of the ORBIT radio grid testbed for evaluation of next-generation wireless network protocols. In: Proceedings of the IEEE wireless communications and networking conference
13. Werner-Allen G, Swieskowski P, Welsh M (2005) Motelab: A wireless sensor network testbed. In: Proceedings of the fourth international conference on information processing in sensor networks (IPSN'05), special track on platform tools and design methods for network embedded sensors (SPOTS)
14. Aschenbruck N, Bauer J, Ernst R, Fuchs C, Kirchhoff J (2011) Demo abstract: a Mesh-based command and control sensing system for public safety scenarios. In: Proceedings of the 9th ACM conference on embedded networked sensor systems
15. Aschenbruck N, Bauer J, Ernst R, Fuchs C, Kirchhoff J (2011) Deploying a mesh-based command and control sensing system in a disaster area maneuver. In: Proceedings of the 9th ACM conference on embedded networked sensor systems
16. Lee SJ, Gerla M, Chiang CC (1999) On-demand multicast routing protocol. In: Proceedings of the IEEE wireless communications and networking conference (WCNC), pp 1298–1302
17. Aschenbruck N, Gerhards-Padilla E, Martini P (2009) Modeling mobility in disaster area scenarios. Elsevier Perform Eval Spec Issue Perform Eval Wireless Ad Hoc Sens Ubiquitous Netw 66(12):773–790
18. Kohler E, Morris R, Chen B, Jannotti J, Kaashoek FM (2000) The click modular router. ACM Trans Comput Syst 18(3):263–297
19. Aschenbruck N, Fuchs C (2011) STMP—sensor data transmission and management protocol. In: Proceedings of the 36th IEEE conference on local computer networks
20. Govaers F, Fuchs C, Aschenbruck N (2010) Evaluation of network effects on the Kalman filter and accumulated state density filter. In: Proceedings of the 2nd international workshop on cognitive information processing
21. Bar-Shalom Y, Li XR, Kirubarajan T (2001) Estimation with applications to tracking and navigation. Wiley, New York
22. Koch W (2009) On accumulated state densities with applications to out-of-sequence measurement processing. In: Proceedings of the 12th ISIF international conference on information fusion
23. Hill J, Szewczyk R, Woo A, Hollar S, Culler D, Pister K (2000) System architecture directions for networked sensors. In: Proceedings of the 9th international conference on architectural support for programming languages and operating systems, pp 93–104. doi:10.1145/378993.379006
24. Costa P, Mottola L, Murphy AL, Picco GP (2007) Programming wireless sensor networks with the TeenyLIME middleware. In: Proceedings of the 8th ACM/USENIX international middleware conference
25. Ceriotti M, Mottola L, Picco GP, Murphy AL, Guna S, Corrà M, Pozzi M, Zonta D, Zanon P (2009) Monitoring heritage buildings with wireless sensor networks: the Torre Aquila deployment. In: Proceedings of the 8th international conference on information processing in sensor networks (IPSN)
26. Mottola L, Picco G, Ceriotti M, Guna S, Murphy A (2010) Not all wireless sensor networks are created equal: a comparative study on tunnels. ACM Trans Sens Netw (TOSN) 7(2):149–162
27. Raza U, Camerra A, Murphy A, Palpanas T, Picco G (2012) What does model-driven data acquisition really achieve in wireless sensor networks? In: Proceedings of the 10th IEEE international conference on pervasive computing and communications (PerCom 2005), IEEE computer society

Chapter 3
Mobility of Cooperating Objects

3.1 Overview

Mobility is a key issue in Cooperating Objects. It has deep influences on the cooperation between objects and affects their main sensing, actuation and communication capabilities. Many factors should be considered when designing mobility management mechanisms for Cooperating Objects such as velocity, obstacles, radio propagation models, network scale, density and partitioning, among many others. Supporting node mobility should not have a negative impact on other QoS metrics, as application requirements should always be respected. Mobility management involves issues related to control, trajectory following and planning, mapping, just to name a few. Mobile Cooperating Objects interacting in the same physical space, involves safety issues related to coordination and cooperation paradigms. Furthermore, mobility requires high levels of security to prevent consequences of malicious attacks.

The complementarity between mobile objects, or between static and mobile objects, is of high interest in a wide range of applications. This chapter briefly presents five examples, in which mobility is an intrinsic component of the cooperation. These applications were selected to represent different sectors and scenarios, ranging from industrial applications to air traffic management and civil protection management. They intend to illustrate the wide range of problems that can be tackled with Cooperating Objects. The presented applications were also selected to represent different levels of maturity in the development of technologies. While industrial applications have advanced maturity, civil protection applications are still object of Research and Development projects although important steps towards their marketing are being carried out.

This chapter presents applications in the following sectors:

- Industrial scenarios
- Air Traffic Management (ATM)

Contributors of this chapter include: Enrique Casado, Gianluca Dini, José Ramiro Martínez-de Dios, Simone Martini, Luis Merino, Aníbal Ollero, Lucia Pallottino, and José Pinto.

- Ocean scenarios
- Urban scenarios
- Civil protection in disaster scenarios.

Present manufacturing and production industries are facing challenging complexity specifications with an unprecedented level of flexibility. This requires a paradigm shift in the organization of the production sites and of the logistic areas. In this context, cooperation of multi-vehicle robotic systems provides competitive advantages with respect to the single-agent solutions in terms of task speed-up, robustness and scalability.

The rapid increase of the air traffic requires a refurbishment of the Air Traffic Management (ATM) paradigm. The current system relies on a clearance-based airspace management, where the demand and the conflict detection and resolution activities are performed tactically by the air navigation authority. The new paradigm adopts a cooperative approach, in which the operations have the freedom to select their path and speed in real time. The similarities between the new paradigm and Cooperating Objects approach are illustrated in this chapter.

The oceans cover two thirds of our planet and are abundant with life, mysteries and natural resources. A recent trend in ocean exploration consists of using collaboration between robots and moored sensors to monitor and reason about surface and underwater phenomena, as well as making this data accessible to scientists in land. Robots can carry sensor payloads and survey large areas in a single run. Moored sensors may remain underwater collecting data for large periods of time. This chapter presents cooperation between robots and static sensors for ocean exploration.

There is a trend in developing intelligent neighbourhoods that extensively use information and communication technologies to offer new services to inhabitants. Cooperating Objects such as robots cooperating with static sensor networks are expected to play a very important role in this type of scenarios. This chapter presents person guidance applications in urban environments by a team constituted by robots and sensor networks.

The cooperation between mobile and static objects has wide application fields also in civil security and protection. The cooperation of ground Wireless Sensor Network (WSN) and mobile sensors on board Unmanned Aerial Systems (UAS) and/or Unmanned Ground Vehicles (UGVs) is very valuable in tasks such as search and rescue of victims, monitoring safety conditions in the disaster, tracking and improving the safety of first responders, deployment of infrastructure etc.

3.2 Mobility in Industrial Scenarios

3.2.1 Overview

Present manufacturing and production industries are facing challenging complexity specifications that require very high levels of productivity and of quality to be matched by an unprecedented level of flexibility and sustainability along with a

Fig. 3.1 An autonomous production plant where industrial robotic systems (such as Laser Guided Vehicles) are largely used and highly integrated with the information systems of the enterprise and of the suppliers

strong reduction in maintenance and reconfiguration costs. This not only requires a paradigm shift in the organization of the production sites and of the logistic areas, but also a much further exploitation of industrial robotic systems.

In this context, the traditional operational scenario where robots, segregated in specific areas of the plants, carry out repetitive and elementary tasks and are controlled by a centralized intelligence is may be doomed to a quick obsolescence for its unexpected inability to guarantee a sufficient level of scalability, robustness and reconfigurability.

Although many recent academic achievements have evidenced a spectacular growth in robot abilities to operate autonomously and in coordinated teams, thus disclosing new opportunities e.g., in disaster management, planetary explorations, surveillance and control on a geographic scale, most of these results remain concealed into labs and do not contribute to the development of competitive technologies for the industry (Fig. 3.1).

Multi-vehicle robotic systems are largely used in industrial transportation and logistics systems, as they provide competitive advantage versus the single-agent solutions in terms of task speed-up, robustness and scalability. For instance, a typical function of a multi Laser Guided Vehicle (LGV) system consists of transporting raw or semi-finished material from a factory's warehouse to its production lines. However, their adoption raises management and coordination problems, such as collision avoidance, conflict resolution requiring fast and reliable negotiation of shared resources. Conflict resolution in the use of these resources is critical for safety and robustness of the operations. Avoidance/resolution of stall situations as well as fluent navigation of the robotic agents must be guaranteed for system efficiency. Stall situations occur when agents are unable to move from the particular configuration (i.e., deadlock) or constrained to move along a finite number of paths without reaching the final destination (i.e., livelock).

In this topic, the academic literature is divided into two categories: centralized and decentralized, see e.g., [1–4]. A method using the notion of composite robot is presented in [5]. Other centralized approaches, using e.g., the *master-slave* control, are proposed in [6] and, using the so-called coordination diagram is presented in [7, 8]. A distributed route planning method for multiple mobile robots that uses so-called Lagrangian decomposition technique is presented in [9]; in [10] authors present a coordination algorithm, which can be considered between centralized and decoupled planning. In [11] a framework for decentralized and parallel coordination system, based on dynamic assignment of robot motion priorities is developed. In that framework only the collision avoidance problem has been addressed, while in [12, 13], the workspace is decomposed into discrete spatial resources and robots move on pre-planned paths applying the concept of distributed mutual exclusion [14] to coordinate their motions.

Many others focus also on the important aspects of collision avoidance and deadlock avoidance, see e.g., [15–18]. In [18] a novel paradigm for conflict resolution in multi-vehicle traffic systems is depicted, where a number of mobile agents move freely in a finite area following a pre-defined motion profile. The key idea is related to the tessellation of the underlying motion area in a finite number of cells. These cells are considered as resources that have to be acquired by the mobile agents for the execution of their motion profile, according to an appropriate resource allocation protocol. The developed protocol is based on the real-time management of sequential resource allocation systems (RASs) and it is able to formally guarantee the safe operation of the underlying traffic system, while remaining scalable with respect to the number of the involved agents. It is worth noting that this approach is applicable even to those traffic systems where all vehicles have to be in perpetual motion until their retirement.

In [19] a method for coordinating the independently planned trajectories of multiple mobile robots to avoid collisions and deadlocks is described. Whenever the distance between two robots drops below a certain value, they exchange information about their planned trajectories. If a possible collision is detected, they start to monitor their movements and, if necessary, they may insert idle times between certain segments of their trajectories in order to avoid collisions. Deadlocks between two or more robots may occur if some robots are blocking each other and none is able to continue with its trajectory without a collision. The method allows these deadlocks to be reliably detected and in such case the trajectory planner of each of the involved robots is asked to plan an alternative trajectory until the deadlock is resolved.

In [20] an architecture that enables multiple robots to explicitly coordinate actions at multiple levels of abstraction is presented. In particular, the authors develop an extension to the traditional three-layered architecture that allows robots to interact directly at each layer. At the behavioural level, robots create distributed control loops, at the execution level they synchronize task execution, and at the planning level they use market-based techniques to assign tasks, form teams, and allocate resources. Each robot uses a complete three-layered architecture, so each can act independently, and if needed it can also coordinate motion with other agents. By allowing each layer to interact directly with its peers, robots are able to create distributed feedback loops, operating at different abstraction levels. In this way, problems that may arise, can be dealt with at the appropriate level, without involving higher layers, so that the latency in the systems is decreased while the robustness of the whole system increases. Furthermore, in [21] and [22] a technique, based on a Petri net, that avoids deadlocks through re-routing is presented. Other strategies, e.g., [23–26], avoid deadlocks by detecting a cyclic-waiting situation, using graph theory for planning paths such that deadlock is a priori avoided or using a matrix-based deadlock detection algorithm.

More recently, in [27] a decentralized path-planning algorithm providing collision-free policy for a group of autonomous agents is proposed. Within this technique agents negotiate their paths via wireless communication. Conflicts between agents are resolved by a cost-based negotiation process, with a handshaking procedure which guarantees agents to be updated with the most recent information about the system. The algorithm can be extended with the introduction of waypoints, which increases performance at the cost of an additional wireless traffic. In [28] some interesting aspects related to the Kiva system are described, which creates a new paradigm for pick-pack-and-ship warehouses that significantly improves worker productivity using movable storage shelves that can be lifted by small, autonomous robots. Although the overall system is cooperative, Kiva robots are essentially independent. No robot depends upon any other robot to accomplish its task, but the system requires them all to successfully complete a customer order. Each robot and station are represented in the system by a drive unit agent (DUA) and by an inventory station agent (ISA), respectively. Robots receive requests and act to accomplish them. At the same time, the system embodies a massive, real-time, resource allocation problem together with resource allocation, task, path, and motion planning by using a control stack with standard abstraction layers.

In [29] the safety of the planned paths of autonomous vehicles with respect to the movement of other traffic participants is considered. Therefore, the stochastic occupancy of the road by other vehicles is predicted. The prediction considers uncertainties originating from the measurements and the possible behaviours of other traffic participants. In addition, the interaction of traffic participants, as well as the limitation of driving manoeuvres due to the road geometry, are considered. The result is the probability of a crash for a specific trajectory of an autonomous vehicle. The presented approach results to be efficient since most of the intensive computation is performed off-line. This data is then used to improve the performance of the on-line algorithm that can be efficiently used for real-time applications.

In [30] an adaptive path planning algorithm for vehicles moving on a grid is presented. It considers a workspace that consists of a symmetric grid and a large number of vehicles that move in the grid to accomplish a certain task. In this approach each vehicle is assigned with the task of visiting a set of randomly selected locations, which are updated over time. The dynamics of the vehicles are described by a constrained linear double-integrator model and the objective is to find in real-time a set of trajectories that maximize the average speed of the vehicles, while ensuring safety through a space reservation mechanism. The trajectory optimization problem is solved locally, whereas a central entity is employed for distribution of information.

In [31] the application of a formal hybrid control approach to design semi-autonomous multi-vehicle systems that are guaranteed to be safe is illustrated. It is proved that, in a structured task, such as driving, simple human-decision models can be effectively learned and employed in a feedback control system, allowing the control to guarantee safety specifications. Deterministic models are here considered even if human decision models are more naturally captured by stochastic frameworks, in which uncertainty due to variability in both subjects and realizations of the same decision is probabilistic.

3.2.2 Application Description/Usage Scenarios

We focus on a very representative application domain for industrial automation. The considered industrial case refers to the logistic scenario of industrial plants with operations typical of process industry, and stores, where the transport of material takes place: away from the endpoint of the processing lines that feed the store, into a temporary storage location, and then away from this and into the start point of the next processing segment. Thus, the material movement takes place at stores, which is usually (but not necessarily) organized with asynchronous operation under performance-based algorithms that pursue optimization based on queue capacity. Briefly, from a conceptual point of view, several options of control distribution may be implemented with increasing level of decentralization:

- A decision maker (or a controller, agent or node of the network) is responsible for a single task
- Any agent is able to manage the task but has only partial information on the task: in this case collaboration must be achieved among agents in order to complete the task.

Centralized control policies are certainly better in terms of trajectory optimality, but they tend to be much more conservative than needed, i.e., robots are assigned with paths that temporally minimize their intersection. Moreover, they are hardly limited by the computational time requirements that increase with the number of robots that are involved. Another disadvantage of centralized control policies is that if the central control unit fails, the whole system is out of control. The major benefits of the decentralized distributed approach can be identified as follows:

- Scalability: ability to handle growing amount of work without compromising efficiency or without the need of reconfiguring the already active agents (and the technology associated)
- Modularity: ability to handle complex processes through the cooperation of simpler agents, allowing faster development from planning and design to production
- Resiliency and fault tolerance: ability of the system to continue to operate correctly even though one or more of its components are malfunctioning or corrupted
- Maintenance and programming: each distributed agent is easier to maintain, program and reconfigure, since interaction with the rest of the plant is minimized

- Hardware reduction: the capability of a common network to manage the communication from and to all the layers reduces network complexity and hardware requirements.

Distributed control implies that a given mission has to be accomplished coordinating the efforts of multiple LGVs, either all equivalent or specialized, with none of them mastering all the other agents. Decisions as how to accomplish the mission are taken not by a single, but by several, cooperating LGVs.

Each of these robots may be aware of the whole mission, of the set of resources available, and of the other cooperating robots. In this case the cooperation is provided consciously by each of them, i.e., it is explicitly coded in their internal software. Alternatively, each robot may be not aware of the mission as a whole and not aware of the other robots: in this case cooperation is provided by the agents not consciously, but compulsorily, i.e., it is implicit in their algorithms.

To give an example, when a shrink wrapper has a pallet ready and calls a LGV to take it away, conscious cooperation implies that all LGVs are aware of this new task, of the other pending calls and their priorities, of their own placement, status, and capabilities. They all negotiate and agree between them who shall go and serve this call, all being aware that the agreed decision is the best possible. On the contrary, a fixed, hard coded decision scheme as "the closest idle LGV will always go" would implement unconscious cooperation.

Within the logistics area, LGVs deliver material from the output queue of the production lines to the input queue of stretch wrapper machines, and are connected to the communication network through a wireless network. In industrial plants LGVs can move along fixed routes consisting of intersections and passage segments. Such areas usually have finite capacity and can be considered as resources to be shared among the LGVs. As the industrial layout may change (also temporally due to obstacles in the environment), the centralized planning must be recomputed with high computational costs, while the system should be shut-off. Decentralized control policies partially resolve the issues of high computational costs in two phases: first, for each agent an optimum path can be defined according to some cost index; then each agent and its neighbours can resolve locally and by themselves conflicts that could arise, according to a shared coordination policy. Notwithstanding this, fully decentralized approaches tend to disregard information that is made available by the infrastructure of the factory, and may turn out not to be the most efficient solutions. These are typically organized into two phases: during a first planning phase robots' paths are computed by using independent objectives for each robot; during a second coordination phase, robots cooperatively manage their motion based on their local neighbourhood situation.

Within this context, research aims at going beyond the state-of-art by seeking the correct trade-off between decentralization and centralization for industrial robot coordination. To achieve this, it applies rule-based, open control policies [1], ensuring conflict avoidance by design. Moreover, robotic agents share a set of rules, that specify what actions they are allowed to perform in the pursuit of their individual goals. Rules are distributed, i.e., they can be evaluated based only on the state of the individual robot, and on information that can be sensed directly or through communication with immediate neighbours [32]. A robot that does not follow the rules due to spontaneous failure or malicious tampering can be considered as an intruder. Intrusion or misbehaviour detection can be detected by cooperation between neighbour robots that can observe the congruence with the rules.

In a decentralized approach a crucial aspect is represented by reliable and secure dissemination of information obtained with local and partial observations of the system's state. If dissemination is unreliable, neighbouring LGVs may achieve inconsistent local views of the system and consequently take inconsistent actions. If dissemination is not secure, an adversary may modify or inject fake messages which may cause Autonomous Guiding Vehicles (AGVs) to achieve wrong and/or inconsistent views. In both cases the coordination task may fail, which raises safety issues. Therefore a reliable and secure state information exchange among neighbours is necessary.

Intuitively, neighbourhood monitoring is crucial for reliable state dissemination because, when an AGV broadcasts its state, an accurate and timely notion of its neighbourhood allows it to track which neighbours have received such state and which have not and thus need a re-transmission. Furthermore, the type of information to be elaborated can be very diverse, ranging from scalar values representing, e.g., proximity measures, to complex sets, representing e.g., the segments or areas that are currently occupied by unexpected obstacles. To effectively disseminate such types of information off-the-shelf consensus algorithms and protocols [33–35] are inadequate. Research pushes for a theoretical advance by proposing an innovative approach based on set-valued consensus [36] and Boolean consensus [37] protocols, which is also applied to the factory-specific scenario.

3.2.3 Key Results and Lessons Learned

The described approach was validated with field experiments in a very representative industrial case-study, based on the specifications of a real paper production sites. Within these plants a team of LGVs moves paper pallets from production lines, to temporary storage locations near stretch-wrappers, and finally to the warehouse. One of the key results has been the design and implementation of a modular architecture of a distributed controller based on a distributed estimation scheme by which the set of LGVs can gather local information and merge it together to determine the resources allocation status. In this sense, LGVs are provided with the ability to distinguish if every neighbouring LGV is correctly performing the assigned task, i.e., it is following the cooperative protocol. This architecture is based on two components: a monitor that is based on the motion rules and on the LGVs' dynamics, and a defined consensus algorithm by which local views of different local monitors can be combined together. As soon as an incorrect behaviour is detected, an alarm is triggered and the LGVs can undertake an adequate countermeasure.

The development of such distributed architecture involved the following activities:

- Analysis of the operative scenarios
- Assessment of possible LGV behaviours and cooperative rules (including the definition of possible ways of violating the cooperative protocol)
- Definition and development of an adequate software interfaces between LGV and the distributed estimation scheme (involving the definition of what information must be exchanged
- Comparison with the centralized solution.

Activities involved the definition of every possible LGV task as a rule-based behaviour, which enabled the description of possible ways of violating the cooperative rules. Moreover, the definition of a resource negotiation scheme among the LGVs required the implementation of a distributed protocol allowing LGVs to exchange information with neighbours via wireless communication. Finally, validation and comparison of the distributed controller with respect to the centralized solution has shown that the developed distributed architecture gives better performance and allows a timely reaction of the system to the unexpected failure of some LGVs. In particular, validation tests allow us to tackle some issues that may arise in real industrial plants such as cases of non-correct identification of an uncooperative LGV, and problems related to the unreliable communication among LGVs due to the wireless communication channel. In most of these cases the on-the-field tuning of the model parameters based on the experimental results has been enough to overcome these issues.

3.3 Mobility in Air Traffic Management

3.3.1 Overview

The rapid increase of the air traffic that nowadays Europe and U.S. are facing requires a complete refurbishment of the Air Traffic Management (ATM) paradigm. The current system relies on a clearance-based airspace management, where the demand and capacity balance, the conflict detection and resolution activities and separation assurance are performed tactically by the air navigation authority. Thus, future needs, such as the management of the expected threefold traffic in 2020 or the integration of unmanned and autonomous aircraft into the controlled airspace, cannot be achieved without a complete shift of the ATM procedures.

The Trajectory Based Operations (TBO) concept[1] represents the improvement of the current ATM system characterized by the implementation of strategic management of the user preferred trajectories. This new approach for air traffic operations

---

[1] "…a safe and efficient flight operating capability under instrument flight rules (IFR) in which the operations have the freedom to select their path and speed in real time. Air traffic restrictions are only imposed to ensure separation, to preclude exceeding airport capacity, to prevent unauthorized flight through Special Use Airspace (SUA), and to ensure safety of flight."

relies on the capability of providing the means for executing the agreed trajectory and the ability of following such trajectory without major discrepancies. The provision of the appropriate infrastructure is under the responsibility of the Air Navigation Server Provider (ANSP) organizations, while the adequate execution of the trajectory depends on the aircraft and the flight crew.

Trajectory management encompasses the process and procedures that establish how the provision and execution have to be performed. This also includes the roles and responsibilities of all involved actors according to the level of responsibility and the mechanism for trajectory planning, agreeing, updating and reviewing. The Federal Aviation Administration (FAA) has estimated that the application of TBO concepts will report a short-term cumulative benefit until 2018 equivalent to $23 billion due to a reduction in the time delay of 35 % regarding current values. This would imply an associated reduction of fuel consumption about 1.4 billion gallons, and therefore, a decrease in CO_2 emissions of 14 million tons during the mentioned period. Although these figures are promising, the major advantages of the new system are expected to be evident in the long-term. The European Commission estimates that air traffic demand in Europe will grow to approximately 25 million commercial flights yearly by 2050 [38], compared to 9.4 million in 2011, and has established ambitious performance targets for that year, with a maximum delay of one minute per flight, a reduction of 75 % in CO_2 emissions per passenger and kilometre and reduction of 90 % in NO_x emissions and less than one accident per 10 million flights, considering a heterogeneous traffic mix of manned and unmanned aircraft.

One of the main challenges to be addressed is the integration of heterogeneous (manned and unmanned) mobile Cooperating Objects (aircraft) with different performance and equipments in a global system capable of supporting Autonomous Aircraft Operations (AAO) and Collaborative Decision Making (CDM), and ensuring at the same time safe operations with an increasing number of operations. The new ATM paradigm also modifies the allocation of separation assurance responsibilities, moving them from Air Traffic Controllers (ATCO) to the flight crew. This relevant change implies improvements of the on-board situational awareness infrastructure to allow a proper execution of the aircraft self-separation activities. To that aim, the Airborne Separation Assurance System (ASAS) will provide the adequate infrastructure to maintain the minimum separation with surrounding traffic.

3.3.2 Application Description/Usage Scenarios

3.3.2.1 Autonomous Aircraft Operations

One of the anticipated features of the future TBO paradigm is that it will allow for Autonomous Aircraft Operations (AAO) [39] in designed regions of airspace. AAO will involve the coordinated transfer of responsibility for separation assurance from the ground-based Air Traffic Control (ATC) system to the cockpit. It is anticipated that autonomous aircraft will not have to fly fixed routes and procedures. Instead, they will be allowed to fly operator-preferred routes, which could be modified dynamically without ATC clearances. The flight crew of autonomous aircraft may alter its intended route to resolve conflicts with the surrounding autonomous aircraft, as well as to take advantage of favourable winds avoiding weather hazards. The envisaged TBO scheme in AAO has the potential to reduce fuel consumption and flight time, which would bring substantial economic benefits to airlines.

The improvements in Communication, Navigation and Surveillance (CNS) expected in future AAO will lead to a shared responsibility among all the stakeholders, mainly the flight crew, the Airline Operation Centres (AOCs) and ATC, for the safe progress of the aircraft through the airspace. CDM represents the procedures and decision support tools that will enable such operations in such future ATM environment. Under the current paradigm, ATCOs and air traffic flow managers are viewed as a central authority with total responsibility both for short-term safety issues and long-term traffic flow scheduling. However, the new paradigm will distribute these responsibilities according to the level of involvement of each stakeholder along the life cycle of the trajectory from planning to execution.

From an ATC perspective, the fact that flight crews of Autonomous Aircraft will hold full responsibility for separation assurance could contribute to alleviating controllers' workload in certain sectors. AAO could improve safety and efficiency of flight operations in regions where radar-based ATC is not available, for example, over oceans and remote continental areas. Currently, monitoring and control of air traffic in this type of airspace, is based on flight plan data, position estimates, and relayed voice positions reports from the pilots. The introduction of AAO in airspace regions currently under procedural ATC would provide the means for flight crews to be made aware of their surrounding traffic and able to detect and resolve conflicts safely and efficiently with no ATC assistance. This would open the way for a reduction of the separation minima in these airspace regions. Added to this, the removal of the requirement to fly along fixed routes would free up airspace currently unused by aircraft flying under procedural control.

3.3.2.2 Collaborative Decision Making

In the new ATM system, any involved stakeholder (ATCOs, airline operational centres AOCs, flight crew, traffic managers …) should be granted to access all the relevant information that supports their own processes. Collaborative Decision Making (CDM) includes all process, procedures and tools that facilitate this accessibility and enable the proper distribution of the collaborative actions among the participating actors commensurate to the different hierarchy levels. The final aim of CDM is to foster collaboration between airspace users and air navigation authorities in order to achieve a more efficient and safe utilization of the airspace.

Collaborative Routing and Collaborative Information Collection and Distribution [40] are the two major areas where collaborative and cooperative procedures have to be designed and developed in order to reach the expected benefits in terms of efficiency, flexibility and robustness.

Collaborative Routing (CR): It refers to the application of CDM process and procedures to route planning and adjustment (reviewing or updating). The objective of CR is to provide the required information in order to accurately forecast demand on the ATM system, mainly airspace density (current and future) and runways throughput, and therefore, to allow the airspace users to pro-actively answer to unexpected changes. Nowadays, the AOCs are reactive players in the resolution of airspace problems associated to bad weather conditions or modifications of the airspace structure. However, in the future collaborative ATM system, the sharing of the related demand and capacity information among all involved actors will facilitate their participation during the problems resolution. This situation will also improve the benefits of the adopted solutions because, theoretically, it will consider at most the user preferences to find the optimum solution that produces the minimum impact to the global community.[2]

Two fundamental operational needs are required as basis of an effective CR: common situational awareness and collaborative re-routing procedures. Common situational awareness implies robust information sharing among all participants, but specially focused on the tight communication of actual position and planning data between the users and the ANSPs. Collaborative strategies to improve the performance of the system have to be based on a common understanding of the real situation. Improved information exchange, as Collaborative Information Collection and Distribution, will provide the means to establish the framework for having a common situational awareness, where the information can be accessed by any authorized player at any required time. Nevertheless, only a common situational awareness is no sufficient to ensure optimum route amendments due to adverse conditions. Collaborative re-routing procedures consider the user preferences when a modification to the actual plan is required due to any reason. This enables the participation of end-users during the resolution process.

Collaborative Information Collection and Distribution: It is commonly assumed that the more available information, the better collaborative decision making. The key issue is that additional information than the strictly needed can overload individuals with inappropriate and unsolicited data. This situation is especially critical in the ATM environment where the common procedures involve air traffic controllers taking real-time decisions based on the available information. To avoid this problem, CDM will provide the processes and procedures to generate a common situational awareness through the dissemination of the information according to the specific needs of each actor at each time. This is one of the main contributions to the future ATM system, the adequate and effective information sharing. In this context, information is not only related to the actual or predicted aircraft position and intent, but also all kind of information generated or consumed by any ATM player ranging from weather forecast, traffic status, or delay estimation until airport capacity data, user preferences or airspace configuration updates.

---

[2] Fairness and Equity are one of the SESAR (Single European Sky ATM Research) Key Performance Areas.

This capability has to be supported by an infrastructure that can be seen as a middleware where the stakeholders can access or release the relevant information according to the role they have in the ATM processes. The System Wide Information Management (SWIM) represents the implementation of this concept. This system will define different levels of responsibility over the released data, giving the appropriate privileges to data generators and data consumers. A correct collection and dissemination of ATM data will produce a higher quality picture of the wide system, facilitating the knowledge of the real system state and improving the situational awareness of all members of the community. Thus, CDM will enhance the efficiency of the system through an optimization of the decision making process based on the global and robust information accessibility.

3.3.2.3 Airborne Separation Assurance System

Airborne Separation Assurance System (ASAS) can be defined as the equipment, protocols and other aircraft state data, flight crew and ATC procedures which enable the pilot exercise responsibility, in agreed and appropriate circumstances, for separation of one aircraft from one or more aircraft [41]. This new system comprises two broad categories of applications:

• Traffic situational awareness, which is related to the provision of information to the flight crew regarding position, identity, flight status and intentions or trajectory predictions of proximate aircraft.
• Cooperative separations, where the pilot uses ASAS equipment to perform operational procedures that aim at maintaining the defined minimum separation with proximate aircraft. The system could advice the flight crew about alternative routes that reduce the likelihood of envisaged conflict situations with other traffic.

These two applications, although currently still in the research and development stages, are considered as cornerstones of the future ATM [42].

The main objective of ASAS is to provide the autonomy to the pilot to maintain the minimum separation with the surrounding air traffic (Fig. 3.2). This objective will enhance the global system capacity and will provide an improvement of the ATC efficiency due to a reduce in controllers' workload.

Under determined circumstances and providing that adequate tools and procedures are in place, the flight crew will exercise responsibility for complying with an ATC clearance that involves maintaining safe separation from other aircraft. The Review of General Concept of Separation (RGCS) panel of International Civil Aviation Organization (ICAO) distinguishes between two levels of transfer of responsibility for cooperative ASAS applications [43]:

• Limited transfer of responsibility. ATC remains responsible for separation assurance, except in determined circumstances defined in a period of time, a volume of airspace and a level of traffic complexity. In such circumstances, the flight crew could assume the responsibility for separation assurance within the boundaries of an ATC clearance.

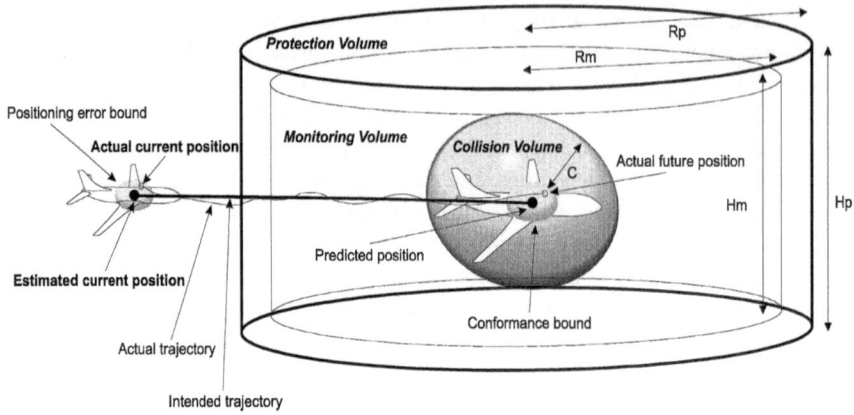

Fig. 3.2 Scheme of the volumes articulating the definition of self separation

- Extended transfer of responsibility. The responsibility for separation assurance is fully assumed by the flight crew. The ground ATM authority would only be responsible for monitoring the traffic complexity and maintaining it at a level compatible with the airborne separation assurance capabilities. This is the case of AAO where an extended transfer of responsibility is applied.

The delegation of separation assurance involves the assignation of specific separation assurance tasks to the flight crew. The RGCS Panel describes the separation assurance process as consisting of four consecutive tasks:

- Conflict detection, which involves the analysis of the traffic situation and the detection of possible violations of the established separation minima between the aircraft considered.
- Determination of conflict resolution strategy.
- Implementation of the conflict resolution strategy.
- Monitoring the conflict resolution strategy.

3.3.3 Key Results and Lessons Learned

The current ATM system is operating with procedures and infrastructures developed since several decades. This system is not capable of handling the increasing requirements coming from the users (airlines) which at the end respond to a social demand for faster and more reliable communications. Due to this situation, two mayor initiatives (SESAR in Europe [44] and NextGen in USA [45]), are trying to define the basis of a new system that overcomes the current limitations and provides a long-term framework to support the foreseen traffic growth. This new system has to ensure the levels of safety are not only maintained but also expanded while improving the efficiency of the operations in terms of fuel consumption or time delays. To do so, the ATM paradigm has to shift to a more autonomous environment where all stakeholders are involved in the decision making process. This new approach will rely on the AAO which establishes the mechanisms that allow the flight crew to take the responsibility for its own flight. This responsibility will be executed in a collaborative manner. Any decisions over any flight will have to consider the implications to other surrounding flights, trying not to affect them negatively. The cooperation among all involved players (mobile objects such as aircraft; or static objects such ground-based systems) is the cornerstone of the new ATM system. Thus, a CDM process has to be clearly defined and implemented based on the capabilities of each system and the improvements to be achieved. This will lead to better performances of the systems because it will be ensured that all users will receive a fair service according to their declared preferences and because any required amendment due to safety considerations will affect the users fairly.

Significant development of advanced systems and infrastructures that support the new ATM paradigm are also needed. These should focus on providing more aircraft autonomy by means of better positioning and information sharing (ADS-B) and by additional capabilities regarding autonomous self-separation (ASAS). A global network infrastructure is also a main advance regarding the current communication system. This network (SWIM) will be used by all systems to have access to the relevant data related to their own purposes. SWIM will also provide all the capabilities that support any cooperative procedure, connecting in real time all the involved actors and showing the information accurate and reliable.

Although some specific areas (such as uncertainty management or ATM information security) have only been lightly addressed and require additional research, clear technical milestones have been defined to reach the goal of having a more efficient and effective ATM system deployed in 2020. Nowadays the industry and institutions (universities and research centres) are focused on designing and developing new system architectures and software applications that support the concepts of AAO and CDM. These future hardware and functionalities are the basis of the automation tools that will support the new ATM paradigm. These tools will connect the stakeholders to the information network and will have the autonomy of taking independent decisions in collaboration with other agents connected to SWIM. This autonomy will reproduce the behaviour encoded during the set-up phase and will facilitate fast and flexible decision making processes.

Cooperative procedures and collaborative decision making are the most relevant features of the future ATM system. All the actors will be aware about any change in the network and any relevant decision will be assumed considering the impact on every single element. Cooperation in the sense of sharing information to facilitate surrounding operations, and collaboration to take the right decision at the right time represent the mayor advances in the ATM field to cope with the envisaged growth of the air traffic in the forthcoming years.

3.4 Mobility in Ocean Scenarios

3.4.1 Overview

The oceans cover two thirds of our planet and are abundant with life, mysteries and natural resources. Phenomena like ocean currents and their slight shifts have a huge economical impact since they dictate global weather evolution and the scarceness or proliferation of natural resources. Still, the oceans remain mostly inaccessible to human explorers due to their vastness and difficult conditions. There are several projects which are now developing sea-floor observatories like NEPTUNE [46] and MARS [47] but these will cover only small fractions of the oceans and are very expensive to deploy.

A recent trend in ocean exploration consists of using collaboration between robots and moored sensors to monitor and reason about surface and underwater phenomena making this data accessible to scientists in land. Robots can carry sensor payloads and survey large areas in a single run. Moored sensors, on the other hand, may remain underwater for large periods of time, collecting data. Data from these sensors is then acquired either using expensive underwater cabling, by retrieving the sensors periodically or by using robots that function as data mules.

Underwater communications are expensive both in terms of hardware costs and power consumption. As a result, robots should work untethered to the surface as possible. Coordination and collaboration between underwater sensors and robots is thus fundamental for ocean monitoring. Wireless underwater communications using optical and acoustic channels require the peers to be within certain distance in order to communicate effectively. Moreover, some sensor payloads like side-scan sonars require the robots to move below certain speed in order to get relevant data. To accelerate surveys one can not speed up the robots doing the survey but instead use multiple robots that divide the survey volume among them to acquire the data faster. This requires underwater communication and collaboration using faulty and bandwidth-limited communications. Hardware and software tools that allow collaboration between operators, heterogeneous robots and sensors are developed at University of Porto, that make it possible to control a network of devices as a whole, exploiting particular advantages of the different devices like sensor types, movement and communication capabilities.

3.4.2 Application Description/Usage Scenarios

There are several applications for networks of heterogeneous robots and sensors, in ocean scenarios. However, here we will focus on three applications for which tangible results exist and are currently framed by ongoing research projects: adaptive sensing, harbour patrolling and underwater mine countermeasures.

3.4.2.1 Adaptive Sensing

Sampling of ocean data over a large area or volume of water may require the use of expensive sensors carried either by manned ships or specialized underwater robots. In order to decrease costs, it is possible to use teams of autonomous robots where part of the fleet will use less expensive sensors (possibly noisy or in some other way related to the variable of interest) and are used to detect areas of interest to be surveyed. After an area of interest is detected, more expensive sensors can be redirected to these areas. This can be done automatically by coordinating different types of robots. Take, for instance, the sensing of underwater chemical plumes. Less expensive Autonomous Underwater Vehicles(AUVs) may carry *pH* sensors while other, more expensive, AUVs may carry water samplers, chemical detection sensors or video cameras.

Multi-vehicle surveys can be optimized by using coordination algorithms such as [48], which is based on the simplex algorithm, using it for searching the minimum of a scalar field using multiple robots. This type of coordination algorithms requires the use of limited communications for synchronization and control of robot behaviour. The Noptilus project (http://www.noptilus-fp7.eu) envisions the use of heterogeneous robots for detection of sources of hazardous material. The project's demonstration scenario comprises the use of multiple AUVs to detect chemical plumes and the use of remotely operated vehicles to take footage and thus, validate potential sources of hazardous material.

3.4.2.2 Harbour Patrolling

Due to the recent evolution of underwater robotics, there is an increased interest in using these technologies for security purposes. One of the applications is the continuous monitoring of critical zones like coastal harbours. This application needs continuous operation of moored sensors and underwater robots, requiring addition and removal of robots at execution time.

3.4.2.3 Mine Countermeasures

There is an increased military interest in the use of underwater robots. Low-cost AUVs can be used to detect and detonate underwater mines at a fraction of the cost of the currently standard methods (mine countermeasure ships). Detection of underwater mines requires the use of side-scan sonars to create sound images of the ocean floor like that in Fig. 3.3 together with on-board computer vision algorithms.

As a result of its collaboration with the Portuguese Navy, University of Porto is building a set of AUVs (Fig. 3.3) in the framework of the SEACON project. The developed vehicles will be used to detect underwater mines autonomously, greatly decreasing the overall cost of this operation. While at the surface, these vehicles communicate with other peers either using GSM or Wi-Fi communications to transfer of gathered data, underwater vehicles use acoustic modems to give state updates and receive simple commands.

Fig. 3.3 Seacon AUVs (*left*) and sound images captured by them (*right*)

3.4.3 Key Results and Lessons Learned

A systems engineering approach to the development of software and hardware tools targeting the creation of networks of autonomous vehicles is being followed. We use a layered approach to planning and execution control that decomposes a complex design problem into a number of sub-problems that are addressed in separate layers, which can be verified in a modular fashion.

3.4.3.1 Control Architecture

Our control architecture [49], depicted in Fig. 3.4, consists of two main layers: multi-vehicle control and vehicle control. Each layer, in turn, is further decomposed into other layers. The vehicle control architecture is standard for all the vehicles. The multi-vehicle control structure is mission dependent. We use vehicle abstractions in multi-vehicle controllers that may reside in some remote locations or in some other vehicles. This leads to different control configurations and strategies.

We developed a software tool-chain composed by on-board software (DUNE), planning and operation software (Neptus) and a common Inter-Module Communication (IMC) protocol. DUNE implements the vehicle control architecture in a predictable and efficient manner for real-time performance. It is used to write generic embedded software by decomposition into asynchronous communicating tasks, which can be reused and distributed as necessary. Tasks communicate between them and with other peers in the network using IMC.

Inter-Module Communication (IMC) [50] is a message-oriented protocol designed for communication between heterogeneous vehicles, sensors, operator consoles and among DUNE tasks. Using the same communication protocol in all software components brings added flexibility so that software components can run on different network locations as long as they use a similar message set (interface). For instance, plan supervisors can run locally at the vehicles or they can run at a central planner or even distributed in the network.

Planning and monitoring is handled by the Neptus command and control infrastructure. This software can be extended and adapted to different vehicles and operations

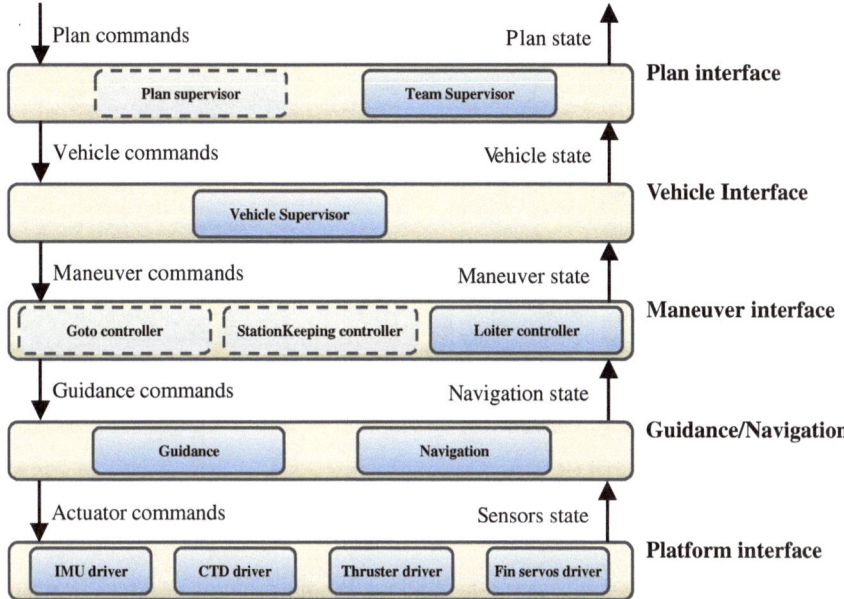

Fig. 3.4 The layered control architecture in use

through a plug-in architecture and provides support for planning, simulation, monitoring and revision of previous missions. In Neptus, plans are a graph of manoeuvres which are units of work to be executed by the vehicles. Transitions between manoeuvres are triggered by conditional expressions (over the variables available to the vehicle). These plans can either be generated manually or automatically based on high-level objectives and parameters like zone to survey and vehicles to be used.

3.4.3.2 Hardware

Different types of robots, e.g., aerial, underwater and surface vehicles, have already been developed, as well as communication and sensing devices like the Manta gateway and stand-alone sensors. In Fig. 3.5 there is a picture with part of the hardware. All vehicles share the same control architecture so they have been combined together for different demonstrations and applications like those we describe next.

3.4.3.3 Vehicle Formations

In order to survey an area faster, multiple vehicles can coordinate their movements so that the areas that are covered are complementary. For this we developed plan generators and multi-vehicle formation manoeuvres. The formation manoeuvre takes

Fig. 3.5 Hardware. **a** NOPTILUS AUV, **b** Swordfish ASV, **c** Seacon AUV, **d** Adamastor ROV,
e Manta Communications Gateway, **f** Raya Oceanography Buoy and **g** Cularis UAV

as parameter a trajectory that multiple vehicles should follow and relative offsets of all vehicles. While executing the manoeuvre, underwater vehicles send manoeuvre completion progress periodically using acoustic communications so that other vehicles may approximate their speed of completion of the trajectory. In 2011 this manoeuvre was demonstrated at Porto Harbour using an underwater vehicle and a surface vehicle that communicated only acoustically.

3.4.3.4 Coordinated Underwater Surveys

In order to speed up underwater surveys, it is possible to use teams of vehicles and divide the area to survey among them. After a survey is finished or in the event of a feature of interest being found by a vehicle, the human operator must be updated as soon as possible. Ideally, the vehicle network is never disconnected, meaning that all nodes are continuously accessible to each other; however achieving this while using acoustic communications is very expensive.

In 2010 a field deployment with an Autonomous Surface Vehicle (ASV) and two autonomous underwater vehicles (AUVs) to demonstrate coordinated surveys using multiple vehicles and operators was realized. The operators on shore generated survey plans by selecting an area to survey and Neptus used a lawn-mowing pattern generator to create plans for all available AUVs. Moreover the ASV was used as a communications gateway, moving to the most centred of the positions of the AUVs whenever a new position was received via acoustic modem. As a result, the ASV was used to increase situational awareness to the operators and to extend the Wi-Fi range of the network.

3.4.3.5 Delay-Tolerant Networking

Unmanned vehicles have limited communications and to counter this, one possibility is to use vehicles' mobility to establish high-latency communication links (use them as data mules between locations). Delay-Tolerant Network (DTN) as defined in RFC4838 [51], is a standardization effort started by the Interplanetary Internet Research Group for creating networking with nodes which are robust to link failures and exploit opportunistic connections to exchange information. The University of Porto has developed a DTN implementation for acoustic modems and has been frequently using DTN in the vehicle deployments.

In 2011 an experiment was conducted where one unmanned aerial plane was successfully used as a data mule to retrieve a file from a submarine which was surfacing at a remote location (inaccessible using wireless communications from the base station). Ongoing work will further test DTN in other scenarios including replanning of vehicle actions and autonomous rendezvous between vehicles for runtime creation of communication links.

3.5 Person Assistance in Urban Scenarios

3.5.1 Overview

There is a trend in developing intelligent neighbourhoods that extensively use ICT technologies to offer new services to inhabitants, like communication facilities, surveillance, etc. For instance, the @22 neighbourhood infrastructure in Barcelona[3] involves new energy, telecommunications, heating and waste collection networks. Sensor networks and Cooperating Objects in general are expected to play a very important role in this kind of applications (see for instance [52, 53]).

Many major cities in Europe are looking for means to reduce the traffic in certain areas, in order to mitigate air and noise pollution, traffic jams, and in general to improve the quality of life. Increasing the amount of pedestrian areas requires that some tasks that are performed nowadays by cars or other mobile platforms, like persons and object transportation, person guidance, surveillance, etc., should be performed by autonomous mobile systems, in general in cooperation with the other sensors and actuators embedded into the infrastructure, like camera networks and other Wireless Sensor Networks.

The key aspects involved are mobility (and its implications into the communications), cooperation, data fusion and human interaction. In this section we will show trials done in the applications of autonomous person guidance by robots and sensor networks in pedestrian scenarios.

---

[3] A Video showing the Barcelona @22 infrastructure is available at http://www.22barcelona.com/content/view/194/609/.

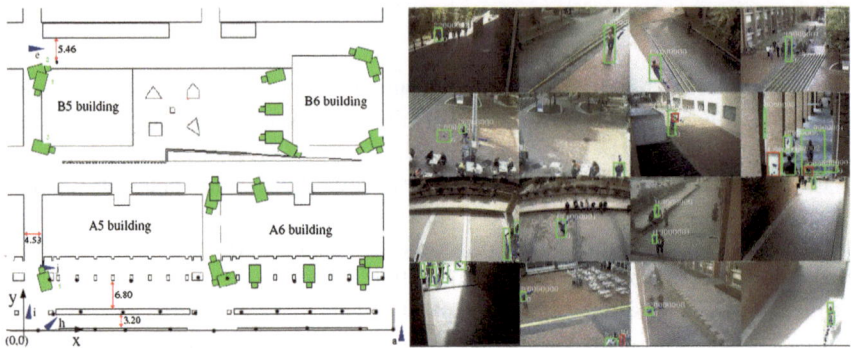

Fig. 3.6 *Left* An schema of the CO network deployed. The full scenario is an square of 100-meter side. The system is composed of 22 cameras, more than 30 Mica2 nodes (*black dots*) and a fleet of robots (not shown). *Right* 16 of the 22 cameras within the experiment system. The cameras can provide overall information about the complete scenario

3.5.2 Application Description/Usage Scenarios

There are many applications that can be considered in a deployment of Cooperating Objects in an urban scenario. Here, we describe one application of person guidance by a team constituted by robots and sensor networks for which real-world deployments have been performed. The idea is that a person, through a mobile phone or other interface, requests the assistance of a robot for guiding him towards a given destination.

The application of person guidance requires the ability to determine and track the position of the person to be guided. This application requires the collaboration of different systems, as, in many cases, a single autonomous entity (i.e., a robot or a static surveillance camera) is not able to acquire all the information required because of the characteristic of the task or the harmful conditions (i.e., loss of visibility). Particularly, this application was tested during the experimental sessions of the URUS EU Project [53]. These experiments were carried out at the Barcelona Robot Lab, which is an outdoor urban experimental robotics site located at the UPC (Universidad Politécnica de Cataluña) campus. A total of 22 fixed colour video cameras were installed and connected through a gigabit Ethernet connection to a computer rack, as well as more than 30 WSN nodes for localization purposes and 9 WLAN antennas with complete area coverage (see Fig. 3.6).

The application illustrates the benefits of having a set of complementary and cooperating sensing objects. The set of fixed cameras can obtain global views of the scene. However, as they are static, they cannot react to non-covered zones, shadows can affect the system and so forth. Robots carry local cameras and can move to adequate places, reacting to the changing conditions. However, their field of view is limited and they can lose the person they are tracking. Wireless devices can also help to localize the people, by estimating their positions measuring the signal strength

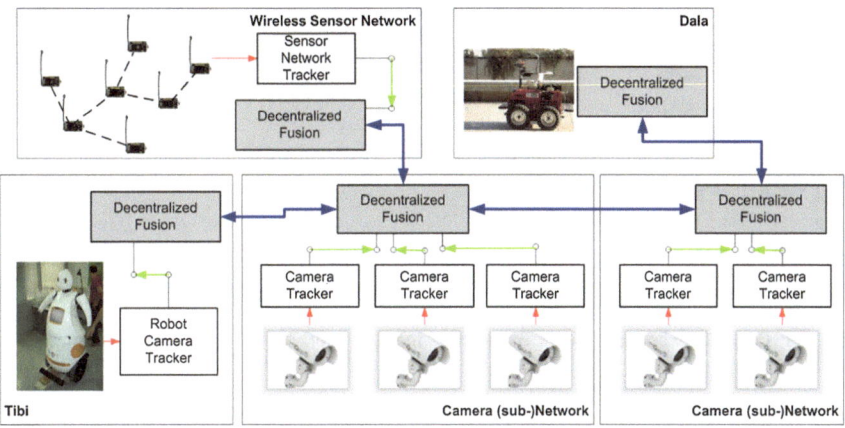

Fig. 3.7 A block description of the URUS perception system. The different subsystems are integrated in a decentralized manner through a set of decentralized data fusion nodes. Locally, each system can process and integrate its data in a central way (like the WSN) or in a distributed way (like the camera network). Some systems could obtain information from the rest of the network even in the case they do not have local sensors

from different static receivers. However, the resolution obtained is usually low, and depends on the density of anchored receivers. Moreover, the application and results obtained show how a set of Cooperating Objects is often a better solution than a single very complex system. The information gathered by the different Cooperating Objects can be fused to improve the performance.

3.5.3 Key Results and Lessons Learned

One of the key results have been the definition of a decentralized data fusion architecture and algorithms in order to scale with the number of sensors in the CO network. Details can be found in [54]. Figure 3.7 shows a simplified version of the system. It consists of a set of fusion nodes which implement a decentralized data fusion algorithm. Each node only employs local information (data from local sensors; for instance, a subset of cameras, or the sensors on board the robot) to obtain a local estimation of the variables of interest (in this case, the position of the person being guided). Then, these nodes share their local estimations among themselves if they are within communication range using wireless links. The main idea is that, as the nodes only use local communications and data, the system is scalable. Also, each node can accumulate information from its local sensors, so temporal communication failures can be tolerated without losing information.

Several experiments of guidance have been conducted in the scenario described above. Figure 3.8 shows some views of the guiding experiments. One important result

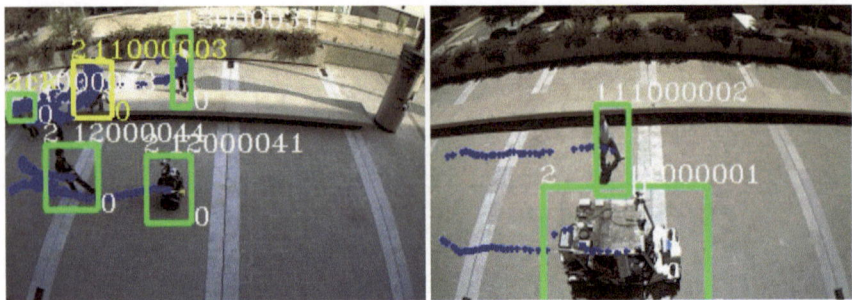

Fig. 3.8 Guiding examples. Views from the camera network of different robots guiding some persons in the scenario

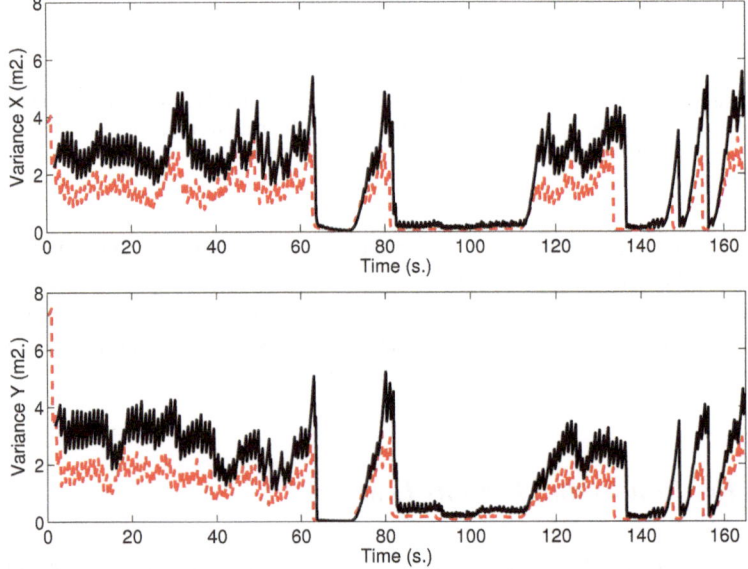

Fig. 3.9 Estimated variance by a central node receiving all the information (*dashed* and *red*) compared to the estimation in a decentralized node (*solid* and *black*)

can be seen in Fig. 3.9. There, the estimation of the decentralized data fusion system is compared to a completely centralized implementation, in which all information is processed in a central node with access to all information at any time. This centralized implementation can obtain an optimal estimation, but it can be a bottleneck of the system and do not scale, so it is not a feasible solution for large teams of Cooperating Objects. It can be seen how the decentralized system obtains estimations close to the ones obtained by the centralized implementations, being at the same time scalable.

3.6 Mobility in Civil Security and Protection

3.6.1 Overview

The cooperation between mobile and static objects opens new fields also in applications related to civil security and protection. In these scenarios the cooperation of ground Wireless Sensor Network, Unmanned Aerial Systems (UAS) and Unmanned Ground Vehicles (UGVs) has high synergies and application potential. For example, firefighters, during fire extinguishing, could deploy WSN nodes equipped with chemical sensors to measure the levels of CO or other toxic gasses produced by fire. Each firefighter could also carry one WSN node with gas concentration sensors in order to monitor the conditions he is exposed to. Multi-hop networks are also valuable to transmit the sensor information between members of the brigade and finally to the vehicles that could be linked to command & control centres. Also, UAS equipped with cameras and other sensors are very valuable to collect information from dangerous and inaccessible locations. A team of UAS could be used in coordination with ground means to locate victims and to help in the rescue. UAS have significant advantages in these scenarios when compared to conventional piloted vehicles such as preventing pilots from flying in dangerous conditions, the ability to fly close to obstacles and the reduction of potential damages in case of accidents.

Cooperating Object technologies could significantly improve the existing emergency management systems. However, despite this large potential market, in the current state of technology the use of Cooperating Objects in civil protection is still object of research and development projects.

A number of projects have been devoted to the integration of robots and sensor networks. Many of them have addressed the development of techniques for the guidance of mobile robots based on the sensor stimuli provided by the wireless sensors, e.g., [55–57]. In [58] mobile nodes are used to increase the quality of service of the static nodes. Coverage, exploration and deployment of sensor nodes have also been proposed [59]. The integration of aerial robots and ground WSN has been also addressed in the last years. Autonomous helicopters have been proposed for WSN deployment and repairing [60]. The ANSER project [61] tackled decentralized data fusion and SLAM using a team of autonomous aerial and ground sensors. There are also relevant projects dealing with multi-robot teams in emergency situations such as the EMBER CMU project aiming at assisting first responders by providing tracking information and the ability for coordination. In EMBER project, range information is used for searching and tracking mobile targets with multiple robots [62]. Multi-agent (combined ground and air) cooperative target localization and tracking has also been demonstrated [63].

Heterogeneity is one of the main requirements in disaster applications. It is not possible to know in advance the constraints in a particular scenario, such as the sensors and actuators that will be needed or the obstacles that will be found. The main approach adopted is to exploit complementarity between heterogeneous objects that cooperate to achieve a common objective. For instance, a robot cannot be equipped with all the sensors that could be useful in a disaster scenario. A cooperative approach in which networked robots equipped with different sensors share and combine their measurements to achieve a common perception is usually adopted.

Avoiding centralization is also important. Besides poor scalability, centralized methods have low robustness, often insufficient to operate in disaster environments usually with harsh conditions. Fully decentralized approaches are commonly adopted.

Self-deployment capabilities are particularly important in disaster scenarios. It allows to substitute malfunctioning components, sensors or communication infrastructure damaged by the disaster. Also, it allows to dynamically adapt the system to the requirements of the situation, e.g., by deploying new sensors to improve monitoring of areas of interest or to repair the network connectivity. In some cases, there is no pre-existing infrastructure at the disaster area. In others, the disaster could have partially or completely destroyed the infrastructure and easiness of integration and extensibility is particularly interesting in order to allow using the components of pre-existing infrastructure remaining after the disaster. This imposes requirements on the system architecture modularity and openness to allow the use of new components.

The EU AWARE project (www.aware-project.net) developed and demonstrated a modular architecture for autonomous distributed cooperation between Unmanned Aerial Systems (UAS), wireless sensor/actuator networks and ground camera networks. One of the main goals of the project was the demonstration of useful actuation capabilities involving multiple ground and aerial robots in the context of civil applications. The project also demonstrated in field experiments the transportation and deployment of loads with a single UAS and also with multiple autonomous aerial vehicles in tight cooperation.

3.6.2 Application Description/Usage Scenarios

AWARE implemented a decentralized architecture for autonomous coordination and cooperation of multiple Unmanned Aerial Systems [64]. Two layers were used for each UAS: the On-board Deliberative Layer (ODL) and the Executive Layer (EL). The ODL was responsible for taking high level decisions in a distributed way. The EL was responsible for the execution of elementary task such as take-off, land, go to a location. The decentralized taking of decisions among the UAS is carried out by a market-based method in which bid messages are interchanged between ODLs of different UAS.

The AWARE platform was comprised of autonomous heterogeneous Unmanned Aerial Systems. The platform was experimentally validated with four TUB-H helicopters developed by Technische Universität Berlin (TUB) and one by Flying-Cam. TUB-H helicopters were equipped with low-weight infrared and visual cameras, devices for accurate deployment of WSN nodes and for lifting and transportation of loads. The Flying-Cam helicopter was equipped with a professional camera and was mainly used to provide high quality images. During the project a new helicopter

Fig. 3.10 WSN nodes deployed during the validation experiments

was developed: the FC III E SARAH (Electric Special Aerial Response Autonomous Helicopter), the first electric helicopter with payload of 7 kg.

The AWARE platform also comprises a network of static camera nodes equipped with infrared and/or visual cameras. The images gathered by the camera network were used in cooperative perception methods. AWARE also comprises a WSN of static and mobile nodes. Two models were used: Crossbow Mica2 and Ambient Series 800. Nodes were equipped with a high variety of heterogeneous sensors including smoke detectors and temperature, humidity and gas concentration sensors useful for detection and monitoring of fire. Figure 3.10 shows some of the WSN nodes deployed in the environment during the validation experiments.

The AWARE platform was equipped with a distributed perception system capable of combining information from the sensors available. A Extended Information Filter (EIF) is used to combine the readings from cameras on-board the UAS, camera network and from the WSN nodes. A fully decentralized delayed-state filter is used to increase the robustness of the estimation against communication delays or sensor failures [54]. This method has been used for tracking of people and for detection and localization of fires. Fire alarms detected by WSN nodes are used to trigger the execution of the mission. Once detected, the method uses visual and infrared images gathered from the camera network or from cameras on-board UAS to confirm the fire detection, estimate its location and measure in real time the fire status. Also, methods based on Particle Filters for localization and tracking using Received (RSSI) have been developed, integrated and validated for tracking of firefighters [65].

Fig. 3.11 Scenario used in the validation of the AWARE project. It was deployed at the Protec-Fire premises at Utrera, Sevilla. The photograph picture shows the structure used to simulate the building

3.6.3 Key Results and Lessons Learned

The full AWARE platform and methods were validated in field experiments in four different missions:

- sensor deployment and fire confirmation with UAS,
- surveillance with multiple UAS,
- tracking of firemen with ground and aerial sensors/cameras,
- load transportation with multiple UAS.

The system validation was carried out in the Protec-Fire (Iturri Group) premises at Utrera (Sevilla, Spain). An urban scenario was simulated using structures as depicted in Fig. 3.11.

These missions included all the functionalities developed within the project. WSN deployment using UAS was used in AWARE as a mean of improving the monitoring capabilities at a certain region of interest. In the experiments, UAS equipped with an infrared camera automatically located the alarm triggered by the WSN nodes. Another UAS equipped with a visual camera identified barrels around the building. Then, the system intensified the monitoring of the area between the fire and the barrels by deploying WSN nodes equipped with temperature nodes. Once deployed, the new nodes integrated in the pre-existing WSN infrastructure and transmitted the measurements to the Monitoring Station. Figure 3.12 shows pictures taken in the experiments.

Fig. 3.12 Automatic deployment of WSN nodes using helicopters

Fig. 3.13 Picture of the fire truck with the implemented fire extinguishing system

Figure 3.13 shows a picture taken of the fire extinguishing system during one experiment. The cooperative perception system measured automatically and in real time the fire characteristics such as location and size using visual and infrared images gathered by the UAS and the camera network nodes. Once, the fire coordinates have been estimated with sufficiently low uncertainty, they are transmitted to the fire extinguishing system. The fire extinguishing system was implemented by a fire truck equipped with an automated water cannon, resulting from the adaptation of a commercial monitor with manual operation. It could be pointed in pitch and jaw angles. Although the driving of the fire truck was done manually, the pointing, activation and deactivation of the water cannon was carried out automatically.

3.7 Conclusions

The complementarity between mobile Cooperating Objects and between mobile and static Cooperating Objects are of high interest in a wide range of applications. This chapter briefly presented some example applications where mobility is an intrinsic component in the cooperation: applications in industrial scenarios, air traffic management, ocean exploration, person assistance in urban scenarios and applications for civil protection in disaster scenarios.

It should be pointed out that although most of the applications presented are still results of research projects, their technological maturity improves at good rate. In some cases, the adoption of the Cooperating Object approach involves normalization and standardization, which is already in process. In others, some products have been already commercialized with high success, which illustrates the good acceptance of Cooperating Object technologies.

References

1. Pallottino L, Scordio VG, Frazzoli E, Bicchi A (2007) Decentralized cooperative policy for conflict resolution in multi-vehicle systems. IEEE Trans Robot 23(6):1170–1183
2. LaValle SM (2006) Planning Algorithms. Cambridge University Press, Cambridge
3. Alami R, Fleury S, Herrb M, Ingrand F, Robert F (1998) Multi-robot cooperation in the martha project. IEEE Robot Autom Mag 5(1):36–47
4. Lygeros J, Godbole D, Sastry S (1998) Verified hybrid controllers for automated vehicles. IEEE Trans Autom Control 43(4):522–539
5. Svestka P, Overmars M (1995) Coordinated motion planning for multiple car-like robots using probabilistic roadmaps. In: Proceedings of IEEE international conference on robotics and automation, vol 2, pp 1631–1636
6. Yuta S, Premvuti S (1992) Coordinating autonomous and centralized decision making to achieve cooperative behaviors between multiple mobile robots. In: Proceedings of the IEEE/RSJ international conference on intelligent robots and systems, pp 1566–1574
7. O'Donnell P, Lozano-Periz T (1989) Deadlock-free and collision-free coordination of two robot manipulators. In: Proceedings of IEEE international conference on robotics and automation, vol 1, pp 484–489
8. Olmi R, Secchi C, Fantuzzi C (2008) Coordination of multiple agvs in an industrial application. In: Proceedings of IEEE international conference on robotics and automation, pp 1916–1921.
9. Guo Y, Parker L (2002) A distributed and optimal motion planning approach for multiple mobile robots. In: Proceedings of IEEE international conference on robotics and automation, vol 3, pp 2612–2619

10. LaValle S, Hutchinson S (1998) Optimal motion planning for multiple robots having independent goals. IEEE Trans Robot Autom 14(6):912–925
11. Azarm K, Schmidt G (1997) Conflict-free motion of multiple mobile robots based on decentralized motion planning and negotiation. In: Proceedings of IEEE international conference on robotics and automation, vol 4, pp 3526–3533
12. Wang J (1995) Operating primitives supporting traffic regulation and control of mobile robots under distributed robotic systems. In: Proceedings of IEEE international conference on robotics and automation, vol 2, pp 1613–1618
13. Wang J, Premvuti S (1995) Distributed traffic regulation and control for multiple autonomous mobile robots operating in discrete space. In: Proceedings of IEEE international conference on robotics and automation, vol 2, pp 1619–1624
14. Lamport L (1986 I and II) The mutual exclusion problem: part i—a theory of interprocess communication and part ii—statement and solutions. J ACM (JACM) 33(2):313–348
15. Roszkowska E, Reveliotis S (2008) On the liveness of guidepath-based, zoned-controlled, dynamically routed, closed traffic systems. IEEE Trans Autom Control 53:1689–1695
16. Fanti M (2002) Event-based controller to avoid deadlock and collisions in zone-control agvs. Int J Prod Res 40(6):1453–1478
17. Reveliotis S, Ferreira P (2002) Deadlock avoidance policies for automated manufacturing cells. IEEE Trans Robot Autom 12(6):845–857
18. Reveliotis SA (2000) Conflict resolution in agv systems. IIE Trans 32:647–659
19. Jager M, Nebel B (2001) Decentralized collision avoidance, deadlock detection, and deadlock resolution for multiple mobile robots. In: Proceedings of IEEE/RSJ international conference on intelligent robots and systems, vol 3, pp 1213–1219
20. Simmons R, Smith T, Dias MB, Goldberg D, Hershberger D, Stentz A, Zlot R (2002) A layered architecture for coordination of mobile robots. In: Proceedings of multi-robot systems: from swarms to intelligent automata, Kluwer, Alphen aan den Rijn
21. Wu N, Zhou M (2007) Shortest routing of bidirectional automated guided vehicles avoiding deadlock and blocking. IEEE/ASME Trans Mechatron 12(1):63–72
22. Fanti M (2002) A deadlock avoidance strategy for agv systems modelled by coloured petri nets. In: Proceedings of 6th international workshop on discrete event systems, pp 61–66
23. Lochana Moorthy R, Hock-Guan W, Wing-Cheong N, Chung-Piaw T (2003) Cyclic deadlock prediction and avoidance for zone-controlled agv system. Int J Prod Econ 83(3):309–324
24. Singhal M (1989) Deadlock detection in distributed systems. Computer 22(11):37–48
25. Yoo J, Sim E, Cao C, Park J (2005) An algorithm for deadlock avoidance in an agv system. Int J Adv Manuf Technol 26(5):659–668
26. Lehmann M, Grunow M, Günther H (2006) Deadlock handling for real-time control of agvs at automated container terminals. OR Spectrum 28(4):631–657
27. Purwin O, D'Andrea R, Lee JW (2008) Theory and implementation of path planning by negotiation for decentralized agents. Robot Auton Syst 56:422–436
28. Wurman PR, D'Andrea R, Mountz M (2007) Coordinating hundreds of cooperative, autonomous vehicles in warehouses. In: Proceedings of the 19th national conference on innovative applications of artificial intelligence, vol 2, AAAI Press, Menlo Park, pp 1752–1759. http://dl.acm.org/citation.cfm?id=1620113.1620125
29. Althoff M, Stursberg O, Buss M (2009) Model-based probabilistic collision detection in autonomous driving. IEEE Trans Intell Trans Syst 10(2):299–310. doi:10.1109/TITS.2009.2018966
30. Roozbehani H, D'Andrea R (2011) Adaptive highways on a grid. Robotics Research, Springer Tracts in Advanced Robotics, vol 70. Springer, Berlin, Heidelberg, pp 661–680
31. Verma R, Vecchio D (2011) Semiautonomous multivehicle safety. IEEE Robot Autom Mag 18(3):44–54
32. Bicchi A, Fagiolini A, Pallottino L (2010) Towards a society of robots. IEEE Robot Autom Mag 17(4):26–36
33. Olfati-Saber R, Fax J, Murray R (2007) Consensus and cooperation in networked multi-agent systems. Proc IEEE 95(1):215

34. Jadbabaie A, Lin J, Morse A (2003) Coordination of groups of mobile autonomous agents using nearest neighbor rules. IEEE Trans Autom Control 48(6):988–1001
35. Fax J, Murray R (2004) Information flow and cooperative control of vehicle formations. IEEE Trans Autom Control 49(9):1465–1476. doi:10.1109/TAC.2004.834433
36. Fagiolini A, Pellinacci M, Valenti G, Dini G, Bicchi A (2008) Consensus-based distributed intrusion detection for multi-robots. In: Proceedings of IEEE interanational conference on robotics and automation, pp 120–127. doi:10.1109/ROBOT.2008.4543196
37. Fagiolini A, Martini S, Dubbini N, Bicchi A (2009) Distributed consensus on boolean information. In: Proceedings of 1st IFAC workshop on estimation and control of networked systems (NecSys), Venice, pp 72–77
38. European Commission (2011) Flightpath 2050: Europe's vision for aviation. http://www.acare4europe.com/docs/Flightpath2050_Final.pdf
39. Advanced air transportation technologies project office (NASA) (2002) DAG-TM concept element 5 en-route free maneuvering operational concept description
40. Michael O, Ball CYC, Hoffman Robert, Vossen T (2012) Collaborative decision making in air traffic management: current and future research directions. Technical Report, National Center of Excellence for Aviation Operations Research (NEXTOR II)
41. Surveillance, WP2 CRSPSWG (1999) Operational and Technical Considerations on ASAS applications. Technical Report, International Civil Aviation Organization (ICAO)
42. Team AP (2002) Principles of operations for the use of ASAS. Technical Report, FAA-Eurocontrol R&D Committee
43. WP5 EP (1998) Assessment of emerging technologies: the specific case of ADS-B/ASAS. Technical Report, European Commission
44. SESAR Consortium (2010) European air traffic management master plan. Technical Report, Single European Sky ATM Research (SESAR). http://www.atmmasterplan.eu
45. Huerta MP (2012) NextGen Implementation Plan. Technical Report, Federal Aviation Administration (FAA). http://www.faa.gov/nextgen/implementation/plan/
46. Delaney J, Heath G, Chave A, Kirkham H, Howe B, Wilcock W, Beauchamp P, Maffei A (2001) NEPTUNE: real-time, long-term ocean and earth studies at the scale of a tectonic plate. In: OCEANS'01. MTS/IEEE Conference and Exhibition, pp 1366–1373
47. Dawe T, Bird L, Talkovic M, Brekke K, Osborne D, Etchemendy S (2005) Operational support of regional cabled observatories the mars facility. In: OCEANS'05. Proceedings of MTS/IEEE, pp 1–6
48. Borges de Sousa J, Johansson KH, Silva J, Speranzon A (2007) A verified hierarchical control architecture for co-ordinated multi-vehicle operations. Int J Adapt Control Signal Process 21(2–3):159–188. doi:10.1002/acs.920
49. Pinto J, Calado P, Braga J, Dias P, Martins R, Marques E (2012) Implementation of a control architecture for networked vehicle systems. In: IFAC workshop—navigation, guidance and control of underwater vehicles, Porto
50. Martins R, Dias P, Marques E, Pinto J, ao Sousa J, Pereira F (2009) IMC: a communication protocol for networked vehicles and sensors. In: Proceedings of the IEEE oceans europe (OCEANS'09). doi:10.1109/OCEANSE.2009.5278245
51. Cerf V, Burleigh S, Hooke A, Torgerson L, Durst R, Scott K, Fall K, Weiss H (2007) Delay-tolerant networking architecture. RFC 4838 (Informational). http://www.ietf.org/rfc/rfc4838.txt
52. Mazzolai B, Mattoli V, Laschi C, Salvini P, Ferri G, Ciaravella G, Dario P (2008) Networked and cooperating robots for urban hygiene: the eu funded dustbot project. In: Proceedings of the 5th international conference on ubiquitous robots and ambient intelligence
53. Sanfeliu A, Andrade-Cetto J, Barbosa M, Bowden R, Capitan J, Corominas A, Gilbert A, Illingworth J, Merino L, Mirats J, Moreno P, Ollero A, Sequeira J, Spaan M (2010) Decentralized sensor fusion for ubiquitous networking robotics in urban areas. Sensors 10:2274–2314
54. Capitán J, Merino L, Caballero F, Ollero A (2011) Decentralized delayed-state information filter (DDSIF): a new approach for cooperative decentralized tracking. Robot Auton Syst 59(6):376–388. doi:10.1016/j.robot.2011.02.001

55. Moore KL, Chen Y, Song Z (2004) Diffusion-based path planning in mobile actuator-sensor networks (mas-net): some preliminary results. In: Proceedings of the SPIE the international society for optical engineering, vol 5421, pp 58–69

56. Batalin M, Sukhatme G, Hattig M (2004) Mobile robot navigation using a sensor network. In: Proceedings of the IEEE international conference on robotics and automation, pp 636–642

57. Yao Z, Gupta K (2010) Distributed roadmaps for robot navigation in sensor networks. In: Proceedings of the IEEE international conference on robotics and automation, pp 3078–3063

58. Bisnik N, Abouzeid A, Isler V (2007) Stochastic event capture using mobile sensors subject to a quality metric. IEEE Trans Robot 23(4):676–692

59. Batalin M, Sukhatme G (2007) The design and analysis of an efficient local algorithm for coverage and exploration based on sensor network deployment. IEEE Trans Robot 23(4):661–675

60. Corke P, Hrabar S, Peterson R, Rus D, Saripalli S, Sukhatme G (2004) Autonomous deployment and repair of a sensor network using an unmanned aerial vehicle. In: Proceedings of the IEEE international conference on robotics and automation, pp 3602–3608

61. Sukkarieh S, Nettleton E, Kim JH, Ridley M, Goktogan A, Durrant-Whyte H (2003) The anser project: data fusion across multiple uninhabited air vehicles. Int J Robot Res 22(7–8):505–539

62. Hollinger G, Singh S, Djugash J, Kehagias A (2009) Efficient multi-robot search for a moving target. Int J Robot Res 22(2):201–219

63. Hsieh M, Chaimowicz L, Cowley A, Grocholsky B, Keller J, Kumar V, Taylor C, Endo Y, Arkin R, Jung B, Wolf D, sukhatme G, MacKenzie D (2007) Adaptive teams of autonomous aerial and ground robots for situational awareness. J Field Robot 24(11):991–1014

64. Maza I, Caballero F, Capitan J, Martínez-deDios J, Ollero A (2011) A distributed architecture for a robotic platform with aerial sensor transportation and self-deployment capabilities. J Field Robot 28(3):303–328

65. Caballero F, Maza I, Merino L, Ollero A (2008) A particle filtering method for wireless sensor network localization with an aerial robot beacon. In: Proceedings of international Conference on robotics and automation (ICRA2008)

Chapter 4
Cooperating Objects in Healthcare Applications

4.1 Overview

Wireless sensor/actuators networks (WSN) have emerged in the recent years as one of the enabling technologies for healthcare applications [1–3] both as body sensor networks (BSNs) and as environmental assistant networks. Still their application in Mobile Healthcare is a challenge because of stringent requirements in terms of reliability, quality of service, privacy and security. In Mobile Healthcare applications, WSN may be coupled with heterogeneous combination of platforms ranging from smartphones, specialized wireless sensing dedicated devices for physiological parameters monitoring, intelligent garments up to small resource constrained smart wireless nodes. Such heterogeneity of resources and constraints requires a wise system-view, where the challenge is to guarantee at one side processing capabilities and flexible interfacing with the external world (e.g., to enable tele-monitoring, tele-rehabilitation and interaction with clinicians and caregivers) and at the other side prolonged lifetime, miniaturization, low-cost and robustness.

Mobile healthcare applications impose strict requirements on end-to-end system *reliability*. Considering that wireless sensing systems for healthcare will be used even at home by patients with motor disabilities and medical staff with little training, loss in quality due to operator misuse is a big concern. Therefore, body worn sensors/actuators need to employ techniques for *self-calibration*, automated data validation and cleansing, and interfaces to facilitate and verify their correct installation. On-board self-calibration and *self-check* abilities to identify system misuse are required. The *robustness* of the measurements must be increased by *sensor fusion techniques* and use of simple, but effective, context information, when possible captured and extracted by the same sensing elements, with the aim of maintaining the system compact and form-factored. Sensor fusion techniques and *context* information

---

Contributors of this chapter include: Davide Brunelli, Elisabetta Farella, Giancarlo Fortino, Roberta Giannantonio, and Raffaele Gravina..

S. Karnouskos et al., *Applications and Markets for Cooperating Objects*,
SpringerBriefs in Cooperating Objects,
DOI: 10.1007/978-3-642-45401-1_4, © The Author(s) 2014

can be of great help also in *power consumption reduction*, particularly thinking to at-home and outdoor use (a main issue not sufficiently addressed and solved in previous researches and projects).

Moreover, the availability of middleware/frameworks for programming BSNs would enable rapid prototyping of mobile healthcare applications based on wearable sensors. Although the literature on these frameworks is still rather short, a few of them have been proposed to date. One of the most relevant, and probably the first attempt to define a general platform able to support various WBSN applications is CodeBlue [4]. It was designed to address a wide range of medical scenarios, such as monitoring patients in hospitals or victims of a disaster scene, where both patients/victims and doctors/rescuers may move and not necessarily be in direct radio range all the time.

CodeBlue consists of a set of hardware wearable medical sensor nodes and a software framework running on the TinyOS operating system (www.tinyos.net). A more recent example is RehabSPOT [5], a customizable wireless networked body sensor platform for physical rehabilitation. RehabSPOT is built on top of SunSPOT technology [6] from Sun Microsystems. RehabSPOT-based WBSNs run a uniform program on all wearable nodes although they may perform different functions during runtime. The system software is based on client-server architecture. The server program is installed and running on the PC while the client program is installed in the remote nodes.

SPINE (Signal Processing In Node Environment) [7–10] is an application level domain-specific open-source framework [11] for fast prototyping of applications based on WBSNs. SPINE provides support to distributed signal-processing intensive WBSNs applications by a wide set of pre-defined physiological sensors, signal-processing utilities, and flexible data transmission. Furthermore, it has a powerful and well designed modular structure that allows for easy integration of new custom-designed sensor drivers and processing functions, as well as flexible tailoring and customization of what is already supported, to fit specific developer needs.

One of the fundamental ideas behind SPINE is the software components reuse to allow different end-user applications to configure the sensor nodes at runtime based on the application-specific requirements, so that the same embedded code can be used for several applications without re-programming off-line the sensor nodes before switching from an application to another. In addition to the systems presented above, a few general-purpose middlewares for WSNs have been customized to develop health-care applications. Specifically, Titan [12] and MAPS [13] have been adapted to prototype physical activity recognition systems based on wearable motion sensors.

In this chapter we will promote the use of the SPINE framework (both version 1 and version 2 of SPINE) as middleware layer atop which to build mobile healthcare applications based on WSNs fulfilling the aforementioned requirements of effectiveness and efficiency. In particular, we will present several SPINE-based applications in the mobile healthcare domain as well as basic applications patterns based on SPINE2 which are the building blocks of more complex mobile healthcare applications.

Specifically, we present (i) a physical activity monitoring system that reaches an overall average recognition accuracy of 97 % using only two wearable motion sensor nodes, a fewer number than the most relevant works, (ii) a physical energy expenditure system that is able to estimate the calories burnt during daily activities in real-time without assuming fixed orientation of the worn motion sensor, and (iii) an emotional stress detection that relies on a wireless sensor system and a monitoring application that, by means of time-domain heart-rate analysis, provides a stress index using only 10 min of observations.

Furthermore, we show three additional interesting real applications in which the use of SPINE2 aims at obtaining a general behaviour on systems which is generally not achievable without using a framework as intermediate abstract layer.

Each one of these case studies points out a different aspect in which SPINE2 can be successfully exploited:

- energy and power savings: the possibility to operate in-network processing and in-node computation avoiding useless and power hungry communication
- load balancing and redistribution: the WSN coordinator can map different task to each single node according to context, local computational load and amount of energy available
- data aggregation and dissemination: data can easily spread inside the network without caring about lower network protocols and hardware, focusing only on application and algorithm implementation.

4.2 Physical Activity Recognition

4.2.1 Overview

Human activity recognition is of critical importance in the m-Health domain as it is the basic building block for a 24/7 monitoring of assisted livings. It is a necessary tool for monitoring daily activity levels for wellness applications; it may help identifying abnormal heart rate variations, e.g., by correlating the rate variations with the current activity being performed, and it can be applied in highly-interactive computer games, to cite a few scenarios.

A wide range of such application prototypes has been proposed, although most of them have not hit the market yet. Here we present some of the most representative and pioneering research efforts and prototypes.

An innovative physical activity monitoring system is presented in [14]. The system is based on the eWatch, a multi-sensor platform that can be worn in several body positions (such as at the wrist, ankle, waist, trousers pocket). Multiple activities (sitting, standing, walking, ascending and descending stairs, running) are recognized in real-time and stored into the device for later analysis. The in-node classifier algorithm is a decision tree fed with time-domain features extracted on-line from the raw readings of a two-axis accelerometer and a light sensor. Other projects [15–18] aim at recognizing more complex activities, including movements (such as drinking, brushing the teeth, writing), and hand or facial gestures, but they combine data from multiple sensor nodes placed in different body positions rather than using a single multi-sensor unit like the eWatch.

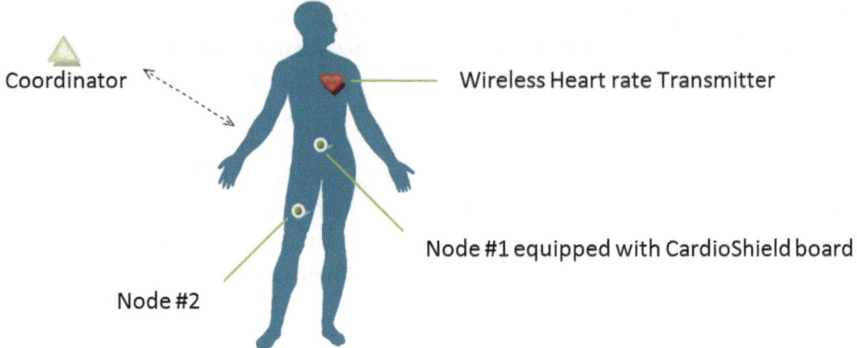

Fig. 4.1 Placement of the sensors on the body

4.2.2 Application Description/Usage Scenarios

The human activity monitoring system prototype here presented is able to recognize postures (*lying*, *sitting*, and *standing still*) and a few movements (*walking*, and *jumping*) of a person; furthermore it can detect if the assisted living has fallen and is unable to stand-up [19–21].

The overall system overview is shown in Fig. 4.1. The wearable nodes are based on the Tmote Sky platform to which is attached a custom sensor-board (SPINE sensor-board) including a 3-axis accelerometer and two 2-axis gyroscopes. The system has been ported on other platforms with different hardware modules and software systems: however using SPINE abstraction APIs the same core application logic, taking care of the acceleration feature data, could be applied to all the solutions. The nodes are powered by a standard 3.7 V, 600 mAh Li-Ion battery). One of them is also featured with a HR wireless sensor; basically, two radios, one 2.4 GHz 802.15.4 and one LF for the HR, coexist on the same board. The wireless solution for the HR monitoring is less invasive then the typical wired electrode based HR monitor.

We also study the heart rate (HR) data for discriminating the intensity of activities. HR may be useful since it correlates with energy expenditure for aerobic exercise; however, if used alone it provides little information about the activity type, in fact it is influenced by other factors for example by emotional states, environment temperature, and fitness level.

The activity recognition system prototype relies on a classifier that takes accelerometer data measured by sensors placed on the waist and on the thigh of the monitored subject and recognizes the movements defined in a training phase. Among the classification algorithms available in the literature, a *K-Nearest Neighbour* [22] (KNN)-based classifier has been selected.

The prototype provides a default training set and a graphical wizard to let the user build his own training set to enhance recognition accuracy. The significant features, that will be eventually activated on the sensor nodes to classify the movements, are selected using an off-line *sequential forward floating selection* (SFFS) [23] algorithm. Experimental results have shown that, given a certain training set, the classification accuracy is not significantly affected by the K value nor by the distance metric used by the classifier. This is because, thanks to the accurate selection of the signal features, the activities instances form clusters that are internally very dense, and well separated among each other. Therefore, the classifier parameters have been selected as follows:

- $K = 1$;
- Metric distance: *Manhattan*.

For the feature selection, the accuracy has been calculated with a shift of 50 % of the data window, using half of the dataset for training and half to test the classifier.

The resultant most significant features are:

- *waist node*: mean on the accelerometer axes XYZ, min value and max value on the accelerometer axis X;
- *leg node*: min value on the accelerometer axis X.

As previously mentioned, the proposed system also includes a *fall detection* module which is implemented on the waist sensor node and can be activated/deactivated at run-time.

Many studies have been done on fall detection using motion sensors and all have to cope with the variety of falls that can occur and the lack of standard datasets to test the algorithm efficiency. Fall detection, in this system, is defined as a rapid modification of the body acceleration (e.g., a crash) followed by a still period in the lying posture. The algorithm has been set and tested empirically with simulated fall situations.

When the fall detector is active, every time that new accelerometer data are acquired, one of the threshold-based functions checks if the *total energy* of the accelerometer signals, exceeds an empirically-evaluated threshold. If so, it sends an alarm message back to the coordinator to inform the user application. False alarms are drastically reduced by a simple mechanism implemented directly at the end-user application: as soon as it receives a fall-detected message, the system waits the recognition of the next seven postures of the person; only if it evaluates four out of seven lying positions, an emergency message is reported to the user attention.

An interesting functionality of the prototype is a simple tool for adding new, user-defined activities among the default ones. The tool drives the user through a simple procedure for acquiring the necessary training data which are then stored in the global data set.

4.2.3 Key Results and Lessons Learned

Although the objective of this prototype concerned mainly in testing the SPINE framework in a semi-realistic use case, the overall performance (see Table 4.1) reached by the recognition system is considerably high, with an average posture/movement classification accuracy of 97 %.

Table 4.1 Posture/Movement recognition accuracy	Sitting	Standing	Lying	Walking	Falling
	96%	92%	98%	94%	100%

The fall detection algorithm is quite accurate as well, as it is able to detected almost every falls, and reaching a low percentage of false alarms.

4.3 Real-Time Physical Energy Expenditure

4.3.1 Overview

Accurate measurements of physical activity are important for obesity research and intervention programs, for fitness and wellness applications, and so on. Physical activity assessment may be used to establish baselines and changes that occur over time. Quantitative assessments may be used to gauge whether recommended levels of regular physical activity are being met, such as 60 and 30 min guidelines for daily moderate intensity activity established by the National Academy of Sciences [24] and U.S. Surgeon General [25], respectively. Conversely, measurements of physical inactivity associated with sedentary lifestyles can equally be useful in estimating risk for overweight and obesity.

4.3.2 Application Description/Usage Scenarios

To provide this quantitative assessment, we developed an energy expenditure algorithm based on previous research conducted by Chen and Sun in which 125 adult subjects wore a 3-axis accelerometer at the waist position for two 24 h periods in a controlled air-tight environment, where whole-room indirect calorimetry was computed based on 1 min measurements of O_2 consumption and CO_2 production [26]. Energy estimates from tri-axial accelerometry were found to be well-correlated (Pearson's r = 0.959) with total energy expenditure measured from the room. The same study used generalized linear and non-linear models to estimate energy expenditure from the raw activity counts along the vertical and horizontal axes.

The problem of many accelerometry-based approaches is an assumption that the sensor is oriented correctly to give accurate activity counts along the relevant axes. This can be difficult to guarantee, particularly for obese subjects, where the sensor may tilt on the waist with activity or changes in posture.

Our Kcal algorithm, instead, improves the current state-of-the-art, by providing a *dynamic compensation of the gravity vector* affecting the accelerometer readings. We first applied time-averaging and vector projection to obtain vertical and horizontal axes regardless of sensor orientation [27]. Briefly, the approach isolates perturbations around a time-averaged (or smoothed) acceleration vector, which indicates the direction of gravity. We compute a vector projection onto this time-averaged representation of gravity to obtain activity counts along the vertical axis, and through a vector subtraction obtain the counts in the horizontal axis. Hence, given the smoothed acceleration vector v that approximates the gravity vector, and an acceleration vector at a given time a, the perturbation is:

$$d = a - v \tag{4.1}$$

The vertical component of this perturbation is computed through vector projection as:

$$p = \left(\frac{d \cdot v}{v \cdot v}\right) \cdot v \tag{4.2}$$

The horizontal vector is the subtraction of p from the perturbation vector d:

$$h = d - p \tag{4.3}$$

Counts along the vertical and horizontal axes, computed as summations of vector magnitude over a period of time, were used as input into the Chen and Sun generalized models [26].

The non-linear equation accounts for variations due to subject weight and gender. Briefly, the equation for energy expenditure EE (in KJ) is based on horizontal and vertical activity counts, H and V for the k-th minute, respectively:

$$EE(k) = aH^{p_1} + bV^{p_2} \tag{4.4}$$

And where a and b are generalized estimates based on the subject's weight w (in Kg) (original equation from [27] differentiates by gender):

$$a = (12.81w + 843.22)/1000 \tag{4.5}$$
$$b = (38.9w + 10.06)/1000 \tag{4.6}$$
$$p_1 = (2.66w + 146.72)/1000 \tag{4.7}$$
$$p_2 = (-3.85w + 968.28)/1000 \tag{4.8}$$

We have implemented the proposed algorithm using the SPINE framework. It partially runs on the sensor node, where we added a new processing functionalities to compute the activity counts, and partially atop the SPINE coordinator, where a graphical application, using the activity counts received every seconds by the sensor node, computes the final estimation of the energy expenditure using the formulas shown above after collecting 1 min of observations.

4.3.3 Key Results

One of the key results is that the wearable device can be arbitrarily oriented on the body, which is particular helpful in real-world usage scenarios, e.g., where the user might place the device randomly inside the trousers pocket. This result has been made possible because accelerometer data are pre-filtered removing the gravity components.

Experiments on 10 subjects walking, running, ascending/descending stairs, and sitting showed high correlation (Pearson's correlation coefficient $= 0.97$) to the results of a commercially available device (www.theactigraph.com), with good relative energy expenditures for different activities; sedentary activity (sitting) producing the lowest energy expenditure, with increasing order of expenditures for walking, stairs, and running.

4.4 Emotional Stress Detection

4.4.1 Overview

The *Heart Rate Variability* (HRV) is based on the analysis of the R-peak to R-peak intervals (*RR-intervals*—RR_i) of the electrocardiogram (ECG) signal in the time and/or frequency domains. Doctors and psychologists are increasingly recognizing the importance of HRV.

A number of studies have demonstrated that patients with anxiety, phobias and post-traumatic stress disorder consistently show lower HRV, even when not exposed to a trauma related prompt. Importantly, this relationship exists independently of age, gender, trait anxiety, cardio-respiratory fitness, heart rate, blood pressure and respiration rate.

This section presents a toolkit based on BSN for the time-domain HRV analysis, named *SPINE-HRV* [28, 29]. The SPINE-HRV is composed of a wearable heart activity monitoring system which continuously acquires the RR-intervals, and a processing application developed using the SPINE framework. The RR-intervals are processed using the SPINE framework at the coordinator through a time-domain analysis of HRV.

The analysis provides seven common parameters known in medical literature to help cardiologists in the diagnosis related to several heart diseases. In particular, SPINE-HRV is applied for stress detection of people during activities in their everyday life.

Monitoring the stress it relevant as many studies show connections between long-term exposure to stress and risk factors for cardiovascular diseases [30, 31].

The main contribution of the proposed system relies in its comfortable wearability, robustness to noise due to body movements and its ability to identify emotional stress in real-time, with no need to rely on off-line analysis.

4.4.2 Application Description/Usage Scenarios

Hardware

The hardware architecture of our system in composed of a wireless chest band, a wireless wearable node and a coordinator station. The wireless chest band detects heart beats and transmits a pulse message over the air each time a heart beat has been detected. It does not require manual power-on nor software configuration. The wearable node is a Telosb mote equipped with a custom board that has a dedicated receiver for the heart beat messages sent by the chest band. Specifically, the wearable node runs the TinyOS operating system and is powered by the SPINE framework. The coordinator station is a PC running a Java application built atop SPINE, which allows bidirectional communication to setup the wearable node and retrieve the heart beats.

Software

The wearable mote runs the SPINE framework, which has been extended with a custom defined processing function to support the custom sensor board. Once enabled by the SPINE coordinator, the processing function on the wearable node starts timestamping the heart beat events, to transmits back to the coordinator the RR_i values.

The RR_i data is used by a Java application built atop SPINE, to compute the average heart beat rate expressed in beat per minute (BPM), the maximum and minimum heart rate, and to analyze the stress level of the monitored subject.

Stress analysis engine

The heart rate is computed from the RR_i values, sent by the wearable node and expressed in milliseconds. It worth noting that we assume a reliable communication between the wireless chest band and the wearable node. The system is able to detect most of the times when heart-beat packets are dropped due to radio interference or out-of-range. We decided not to interpolate dropped RR_i messages to avoid further bias while executing the analysis.

We use a 20-point moving average filter over the inter-beat intervals. Maximum and minimum heart rate, however, are computed instantaneously by dividing the current RR_i received from the wireless node by 1 min.

The stress level of the subject is refreshed every 10 min (previous works have shown that this is the minimum collection time to get significant results [32]). Our approach is based only on a time-domain analysis, which is fair enough to evaluate the stress condition as demonstrated in [32].

Table 4.2 Stress threshold
for HRV parameters

Feature	Threshold	Unit
HR	>85	1/min
pNN50	<7	%
SDNN	<55	ms
RMSSD	<45	ms

Specifically, $\overline{RR_j}$ (computed by averaging on 15 heartbeats) proportional to $\overline{HR}$, *SDNN*, *RMSSD*, and *pNN50* are computed as follows:

$$\overline{RR_j} = \frac{1}{15} \sum_{j=1}^{15} RR_j \tag{4.9}$$

$$SDNN = \sqrt{\frac{1}{N-1} \sum_{j=1}^{N} (RR_j - \overline{RR})^2} \tag{4.10}$$

$$RMSSD = \sqrt{\frac{1}{N-1} \sum_{j=1}^{N-1} (RR_{j+1} - RR_j)^2} \tag{4.11}$$

$$pNN50 = \frac{NN50}{N-1} \times 100 \tag{4.12}$$

where RR_j denotes the value of jth RR interval and N is the total number of successive intervals. *SDNN* is the primary measure used to quantify HRV change, since *SDNN* reflects all the cyclic components responsible for variability in the period of recording. Under negative emotions, the activation of *Autonomic Nervous System* (ANS) is decreased compared to positive emotions; hence, higher SDNN is often an indicator for ANS activation.

The proposed work focuses on determining whether the monitored subject is under emotional stress. It is a *decision problem* that has been solved with a threshold-based approach. Table 4.2 reports the threshold values extracted from the results found in [32]. The final decision is made on a simple majority vote: if three out of the four features exceed the threshold, the current emotional condition is classified as *stressed*.

4.4.3 Key Results

The emotional stress detection relies on a wireless system and a monitoring application that, by means of time-domain heart-rate analysis, provides a stress index using only 10 min of observations. A Key result is related to the detection algorithm that is based uniquely on a time domain analysis, which allows for efficient implementations on embedded devices.

4.5 Physical Rehabilitation

4.5.1 Overview

Wearable wireless sensors might be also used for physical rehabilitation purposes. It is quite common to require repetitive physical exercises for instance to recover from a muscle strain or a surgery. Having a real-time feedback about the exercise performance quality would allow users to independently exercise properly without the need of a continuous professional assistance. Motion sensors are the most appropriate for this kind of applications: they can be worn on the part of the body that needs to be exercised and report precise data about how the movement is being performed. Different motion sensors (accelerometers, gyroscopes, magnetometers, etc.) would provide different information about the position of the body in space.

In [33], a wearable system for knee angle measurement, called KneeMeasurer, is described. It is based on two Bluetooth-enabled sensor node equipped with two-axis accelerometers. Another interesting project is presented in [34]; here, inertial (accelerometer and gyrometer) sensor data are combined with the Microsoft Kinect to estimate ankle and knee angles. Although the results in terms of accuracy are promising, the main drawback of this system is the limited sampling frequency obtained from the Kinect, which is inadequate for faster movements. Finally, a technique for joint angle estimation that uses a combination of rate gyroscope, accelerometer and magnetometer sensor signals is presented in [35]. In this work, the main limitation is due to the errors that can be introduced by the magnetic field distortions commonly found in modern buildings.

In our physical rehabilitation system, we used SPINE wireless nodes equipped with 3-axis accelerometer sensors to monitor in real-time arms and legs movements, with specific attention to both elbow and knee angle measurement.

4.5.2 Application Description/Usage Scenarios

The application consists of monitoring legs and arms bending movements in real-time and comparing them with the ones recorded during set up phase. Main requirements of this application are real-time processing and high precision in movement detection, therefore we did not use SPINE distributed signal processing capabilities but we used its raw data streaming functionality. This way we all sampled data is provided to the SPINE coordinator as input to the algorithm. Despite not using the SPINE on-node signal processing functionalities in this application, the usage of the SPINE framework did speed up the development process.

The standard on-node firmware application (SPINE node side) removed the need to develop any additional nesC code. Additionally, the existent communication framework between the node and server SPINE applications allowed developers to focus on Java implementation of the classification algorithm on the coordinator computer. The application scenario consists of two steps, namely set up and exercise phases. During the set-up phase, the user wears a couple of sensors on either leg or arm that needs to be exercised and performs the correct exercise under the guidance of the rehabilitation professional.

Meanwhile the system records the data and stores it as reference exercise. Set up phase might be optional (as default references are provided) but it is strongly suggested under supervision of the rehabilitation personnel. Then, during the exercise phase, the user repeats the bending movement and is guided with a real-time visual feedback about how the movement is being performed with respect to the stored reference. The application can also provide and store a final score indicating how good the exercise has been performed. Such score can be an indicator for a therapist to assess effectiveness and decide on future course of patient's exercises.

4.5.3 Key Results and Lessons Learned

To evaluate the angle estimation accuracy of our system, we have attached the sensor nodes to the traditional protractor tool used by rehabilitation personnel. We compared the angles estimated by our system against the actual angle obtained from the protractor by changing the protractor arms slowly. The average error is $\pm 1°$, with a maximum error of ± 3 degrees when the movements is done on a fixed plan and can be resolved as a 2D tracking problem, while tracking real time joint angles in the space require the usage of other motion sensors (e.g., magnetometers and gyroscopes). However, we found out that it is important to place the nodes properly and steadily on the arm or leg to avoid motion artefacts. Furthermore, as the current algorithm takes into account the gravity components across the three accelerometer axes, fast movements may introduce significant inertial accelerations that in turns could lead to temporary incorrect angle estimations.

4.6 Energy Aware Fall Detection

4.6.1 Overview

The major contributions to power consumption in sensor nodes are: (i) the power consumed by the digital part for acquisition and elaboration of sensor data and (ii) power for inbound and outbound transmissions. Therefore to apply power-aware techniques we can operate in two different domains: (i) communication protocols and (ii) communication-aware software.

Consider a network where a sensor has to gather data from sensors on which it is necessary to perform some kind of operation. Data analysis can be performed on another node in the same network or on the node itself. These two choices have very

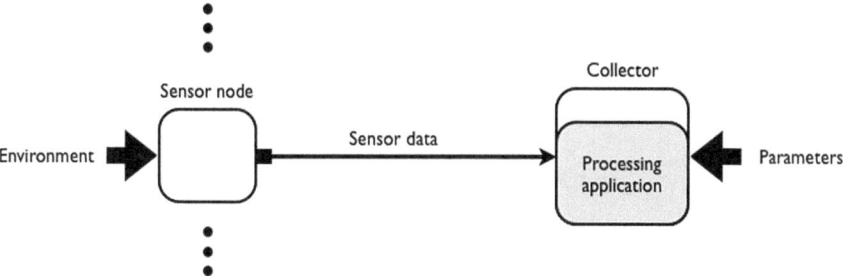

Fig. 4.2 Communication scheme without using a framework. Data are sent back to coordinator without any further elaboration

different impact on energy consumption, involving a different amount of packets sent within the WSN. If digital signal processing is done inside the acquiring node, the output transmission is avoided, saving the energy spent in communication.

According to this new vision the node is not only a passive device that sense and send data to a coordinator but it becomes a true computational unit. The shifting of the core of the analysis from a centralized point to a distributed one creates new problems in coordination and control that cannot be faced without using a new approach. The energy saving obtained with a local computation is compensate by an increasing difficulty in configuration and control over the network, opening a new trade-off between power saving and configuration.

Figure 4.2 presents a case in which a specific algorithm has been implemented over a set of data acquired by a remote sensor node. In this case the node is used as a data forwarder towards central coordinator that can be configured with specific parameters by the user. Since each sample coming from the environment is sent toward the gateway a high amount of energy is wasted in communication, proportionally to sampling period and acquisition time. Moreover with this configuration the coordinator should have enough computational power to serve each request and elaborate each data stream, hence the system is not scalable at all.

Through the use of a specific framework for signal processing in node environment (SPINE2) [20] the allocation of specific task on sensor nodes and the acquisition of the result is a very simple procedure. SPINE2 permits to dynamically allocate and start (but also stop and deallocate) tasks on a specific node, configuring it over-the-air also at run-time. The data effectively sent to the coordinator is only the results of the on-node elaboration (Fig. 4.3).

A very successful clinical application of this framework is the fall detection in which the acceleration is taken as a parameter to monitor detecting falls. Fall detection is one of the hot topic in body area networks, particularly in the ageing society [36]. The happening of the fall event however is unpredictable, therefore BAN must monitor continuously user activity even if transmission can occur only when the fall event is detected. Power consumption must be therefore optimized for a specific scenario: long-term monitoring and sporadic transmissions.

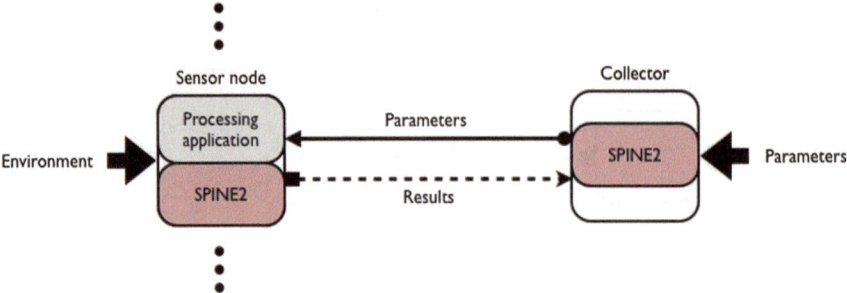

Fig. 4.3 Energy-aware communication scheme using SPINE2. Data processing is done on-node and just the result is sent back

4.6.2 Application Description/Usage Scenarios

A series of nodes equipped with accelerometers are posed on several part of the body with the aim to capture accelerations along three axis: according to the value of a special parametric function a movement is classified as a fall or not. The classification is done by comparing the value of the estimation function with an opportune threshold [37], if the value of acceleration is above the threshold then the movement is classified as a fall. The most common functions used [38] are:

$$SV = \sqrt{(A_x)^2 + (A_y)^2 + (A_y)^2} \tag{4.13}$$

$$Z_2 = \frac{SV_{TOT}^2 - SV_D^2 - G^2}{2G} \tag{4.14}$$

where $A_x(t)$, $A_y(t)$, $A_z(t)$ are the acceleration components along x, y, z axis, G is the gravitational component and SV_D is the vector SV without considering the contribution of gravity. A very common scenario sees several sensor nodes connected over a WSN to a central coordinator, generally a portable device able to run Java and log events. Consider the situation in which a framework is not available: the developer has to implement fixed estimation function with immutable threshold, proceeding to write a new firmware for each change in parameters.

In this context SPINE2 provides an abstract and configurable layer building on it the entire set of function for fall detection, as represented in Fig. 4.4. The coordinator can choose run-time and exchange the function used to evaluate the fall using few configuration packets. Moreover using Java API it is possible to modify value and typology of threshold used to identify the fall. Since this kind of characterization is node oriented, different kind of functions and thresholds can be used for several sensors. This highly dynamic configuration mechanism is extremely power-aware: a single packet of alarm is sent back to coordinator if a fall is detected. Every other data processing is performed on the node using parameters received by coordinator.

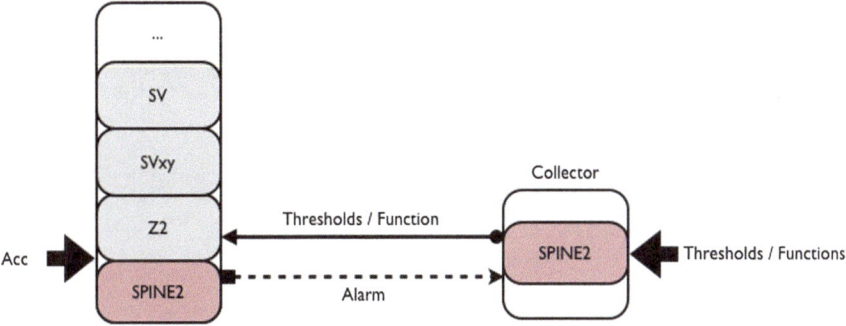

Fig. 4.4 SPINE2 in fall detection application reduces energy consumption of the node. The coordinator run-time can choose and change the evaluation function and value of thresholds as well

4.6.3 Key Results and Lessons Learned

Figure 4.5 presents the oscilloscope trace for a node sending out the alarm packet. The visible spike is related to a peak in power consumption up to 120 mW in correspondence to the packet sent to coordinator. The power consumption is always at minimum level since the radio is always off except when a packet has to be sent and when the node polls the coordinator asking for configuration packets. Since the radio consumption is the major contribution to the overall energy wasting, the use of a framework infrastructure like SPINE2 can improve the overall system by providing energy-awareness and on-the-fly configurability.

4.7 Distributed Digital Signal Processing

4.7.1 Overview

SPINE2 is also suitable framework to exploit the concept of collaborative signal processing where the intrinsic limitation of the nodes can be overcome by a collaborative computational algorithm, such as in [39, 40]. In a complete distributed system without a central unit for control and mapping of tasks, load balancing is a very difficult process since none of the nodes has a fully knowledge of the state of WSN.

SPINE2 can actually implement a fully transparent task mapping, load balancing and data redistribution according to the effective load of the nodes of the WSN. According to recent advances in the field of DDSP (Distributed Digital Signal Processing) an approach in which the load is distributed among more processing units can increase the energy efficiency of the overall system together with an increment

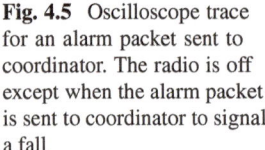
Fig. 4.5 Oscilloscope trace
for an alarm packet sent to
coordinator. The radio is off
except when the alarm packet
is sent to coordinator to signal
a fall

of the performance. Also, when high network reliability is required, redundancy in
collected data and processing is desirable and can be obtained by a redistribution of
acquired data on more than one single node. The idea at the base of DDSP is a to
split the general problem is a series of smaller and simpler sub-problems to map to
different nodes, each working on a smaller data subset.

4.7.2 Application Description/Usage Scenarios

Figure 4.6 presents a general model for DDSP using SPINE2 framework. Data are
collected by one node in the network and partially sent back to the coordinator.
The amount of data sent back is function of the used algorithm. Data gathered by
coordinator are forwarded to other nodes for parallel computation, creating a cluster
of nodes working on a same problem. Data redistribution and diffusion is done
according to the local computational load of each node. The coordinator knows
exactly the tasks allocated and it is able to take proper decisions about which node
include into cluster.

SPINE2 is able to hide the specific implementation of this mechanism to application layer, acting as an invisible omniscient player. The decision about node to
involve in computation is done on-line considering the instantaneous load. One possible application of the theory of DDSP applied on a SPINE2 supported network is
computation of a Fast Fourier Transform (FFT) in a distributed form. The processing
operations can be divided among more sensor node to speed up total process reducing
at same time the energy consumption [41].

Initially for the sake of simplicity we can limit the number of sensor to 2, s_1 and s_2.
If we consider a decimation-in-time FFT, half of sensed data are processed by each

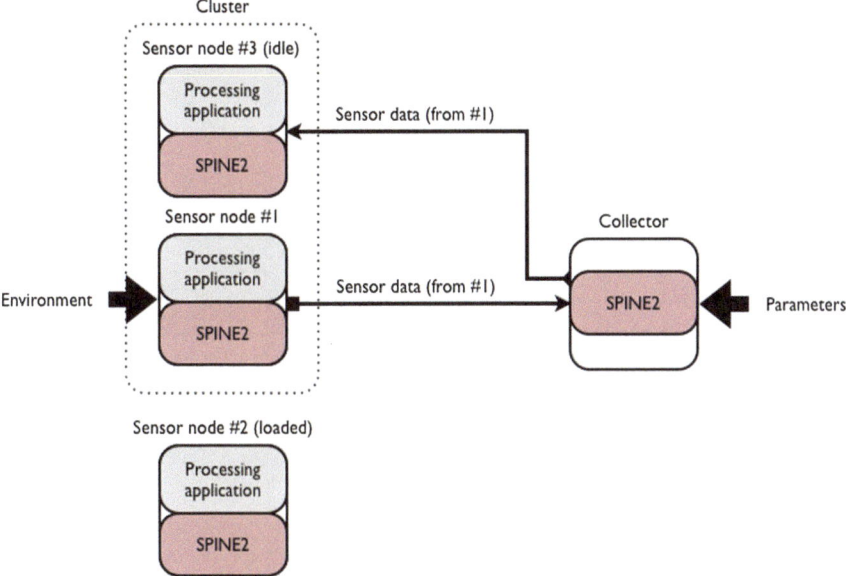

Fig. 4.6 Task mapping, load balancing and data redistribution according to the load on each node. Data coming from node #1 are partially forwarded to node #2 that takes part in processing, creating a cluster of nodes collaborating on a same problem

node. That is if N is the number of sampled acquired by a node, $N/2$ data samples are elaborated by the other one. Named v the sensed vector, this is partitioned into two vector with the same length v_1 and v_2. If w are the weights needed to compute FFT vectors, the algorithm involving SPINE2 is as follows.

1. s_1 acquire N data and send vector v to coordinator.
2. coordinator, according to instantaneous load of each node, chooses the second node to insert into cluster for FFT processing and sent it vector v.
3. s_1 computes $v_1 + w \cdot v_2$.
4. in parallel s_2 estimates $v_1 - w \cdot v_2$.

These operations are totally transparent for s_1 that does not care about position and address of s_2. The use of SPINE2 permits hiding of network topology and structure. The only information needed to each node is relative to its coordinator that acts as a controller and supervisor for the entire operation even without participating to actual data processing.

This described approach is very generic and can be extended to an arbitrary number of processing node. An improved version of this kind of parallel elaboration for DDSP takes into account more than two sensor nodes on which the load is redistributed according to the real computational power available. In most of cases a sensor node is not totally idle but it is busy in some computation, leaving only part of the processor for other tasks. Since it is the coordinator that allocates tasks, it knows

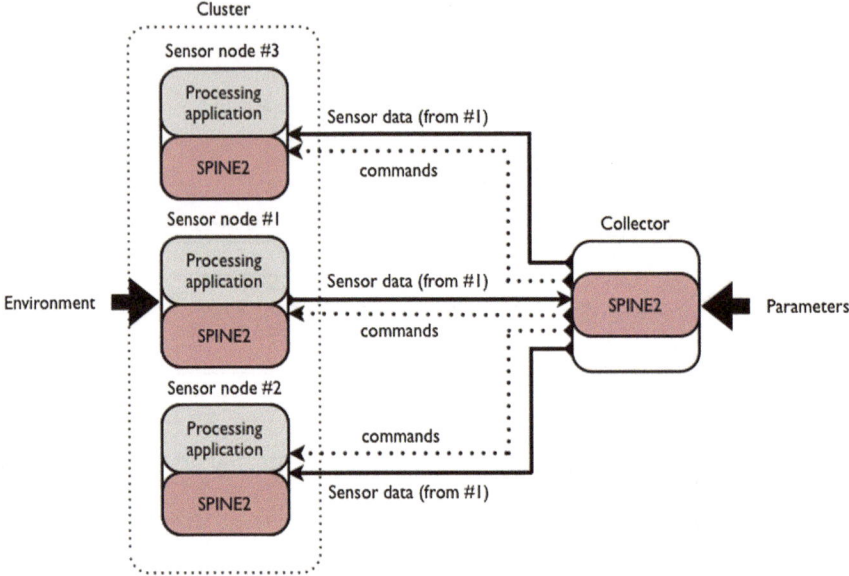

Fig. 4.7 Data gathered from node #1 is sent to coordinator that distributes the vector among nodes in the cluster together with commands regarding the amount of data to process according to the number of tasks already running on those nodes

how much of CPU power is available on each node and therefore it can efficiently allocate a process for FFT computation.

In particular once the coordinator receives the vector data (Fig. 4.7), this is forwarded to nodes that have enough free computational resource. Moreover the vector coordinator gives proper information about the amount of data to process. In this way it can subdivide the work assigning more data to nodes with less previous allocated tasks. The process of dynamic mapping of tasks permits to implement on nodes only DSP algorithm neglecting specific details about coordination and data exchange, simplifying programs development.

4.7.3 Key Results and Lessons Learned

SPINE2 brings additional extra packets and then energy consumption, but the global energy balance is positive and favourable to adoption of the middleware that permits to exploit better packet management policy and save the number of payload byte sent.

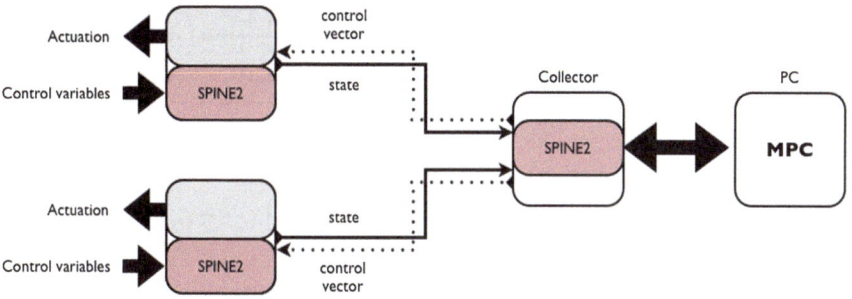

Fig. 4.8 State vector are sent through the coordinator to a PC running a MPC that sends back to nodes the computed control vectors

4.8 Model Predictive Control

4.8.1 Overview

Model Predictive Control (MPC) is a very interesting feedback strategy in which linear models are used to predict the system dynamics even though the dynamics of the closed-loop system is non-linear. The main idea of MPC is to select a control action solving on-line an optimal control problem. The aim is to minimize a specific cost function over a future horizon considering the constrains on the manipulated input and output, where the future behaviour is extracted according to a model of the plant [42].

Practically at a certain time t the current state plant is sampled and the state is used as parameter in a cost function to minimize on-the-fly, obtaining a cost-minimizing control strategy for a short time horizon in the future $[t, t + T]$. After the implementation of the control strategy, the horizon is shifted one step ahead and the algorithm can computes the new strategy.

An application of MPC involves the use of sensor nodes to gather data about the system state, but solving an optimization problem is an hard problem requiring a lot of computational effort and the limited resources of nodes in a WSN are not enough to face such a complexity. For this reason in most cases data containing the state of the controlled system are sent to a device with more computational power able to solve the optimization problem. Once the new vector containing the control policy is computed, this is sent back to the nodes that can perform the control, as presented in Fig. 4.8.

A typical situation is that one in which the two sensor nodes represented are working on the same system or that nodes data are correlated in some way (for example they are in the same environment and performing some kind of measurements on it). This means that the actions performed are not independent one each other and the information about the state on one node is correlated to the state of the other one. In a real case application we can have two nodes that receive their

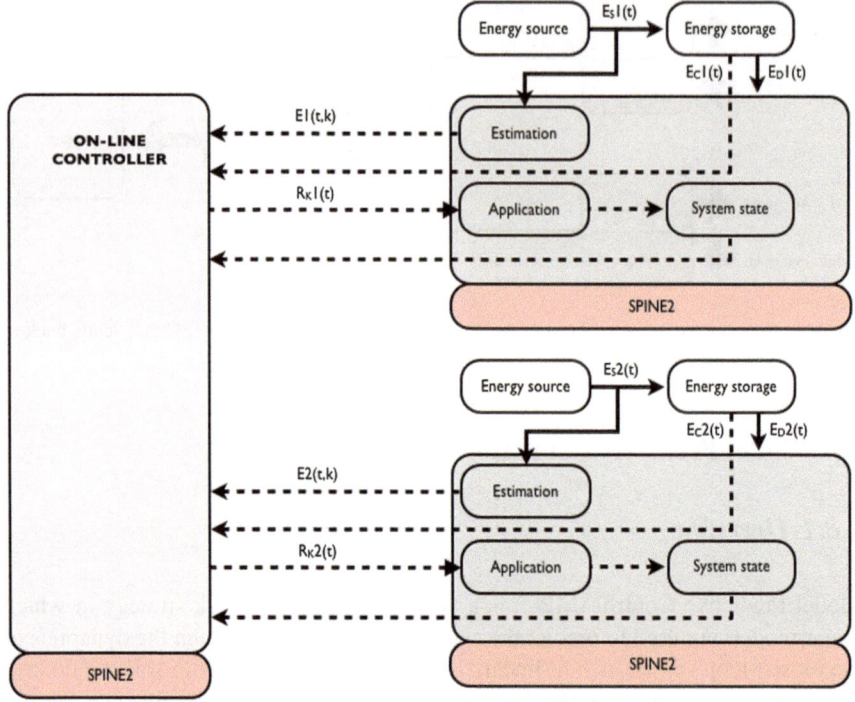

Fig. 4.9 MPC scheme using SPINE2 as underlying framework. Data elaboration is not performed on-node but an external controller process data coming from WSN to extract an optimal solution

energy from solar cells as in [43, 44] or from multiple alternative energy sources such as in [45–47]. In this case during the computation of the best state trajectory it is necessary not only dealing with the minimization of power consumption but also considering the time-varying amount of energy available.

4.8.2 Application Description/Usage Scenarios

The system model is presented in Fig. 4.9 in which two SPINE2 running nodes are powered by energy harvesting devices that at each time t supply energy $E_S1(t)$ and $E_S2(t)$ to the energy storages. At the same temporal instant t part of the energy, $E_D1(t)$ and $E_D2(t)$, is adsorbed by the node, leaving an amount of energy $E_C1(t)$ and $E_C2(t)$ for further use.

On the device there are two main tasks running: the application (depending on the goal of the WSN) and the estimator that predicts future energy production of the harvester, based on past history. The controller is left outside the nodes, for two main reasons: (i) it is computationally the most intensive part and nodes are not able to perform such a heavy calculations on a resource-constrained hardware as an embedded system and (ii) a so centralized controller can consider the performance and state of both sensor node, to develop a better control strategy.

The controller adapts properties of the applications, $R_K1(t)$ and $R_K2(t)$, based on estimation of future available energy, the energy currently stored and the information about the system state, to optimize the overall objective respecting the system constrains. An interesting case is considering the parameters $R_K1(t)$ and $R_K2(t)$ as the sampling rate of sensing tasks, that is a task on node 1 is instantiated R1-times in the interval $[t, t + T)$ and the same for the task on the second node.

For these two nodes we can formulate the following linear program LP:

$$\text{maximize } \lambda \text{ subject to:} \qquad (1)$$

$$\forall 1 \leq i \leq 2 \qquad i = \text{Nodes}$$
$$\lambda \leq s_1(t + k \cdot L) + s_2(t + k \cdot L) \qquad \forall 0 \leq k < N$$
$$s_i(t + k \cdot L) \geq 0 \qquad \forall 0 \leq k < N$$
$$E_{C,i}(t + k \cdot L) = E_{C,i}(t) +$$
$$+ \sum_{j=0}^{k-1} \left(\tilde{E}_i(t, j) - L \cdot s_i(t + j \cdot L) \right) \geq 0 \; \forall 1 \leq k \leq N$$
$$E_{C,i}(t + N \cdot L) \geq E_{C,i}(t) - 100$$

in which $s_i(t)$ is the time-variant rate activation of task, L is the prediction interval and N is the number of intervals in a day.

The solution of the LP problem involves not only a non-negligible computational power but also the knowledge of the parameters of two systems. For this reason the on-line controller cannot be inside the nodes, but it has to be an external system in charge of computing the optimal solution. Even in this case the framework infrastructure of SPINE2 allows to build up the entire system without caring useless details not related to MPC problem. The SPINE2 core on the nodes takes care of communication between node and controller while the gateway is a proxy toward an implementation of resolution algorithm for the LP problem that can resides on the gateway itself.

4.8.3 Key Results and Lessons Learned

SPINE2 makes easy the process of aggregation and dissemination of packets inside the network but the performance of the network itself is very important to ensure the correct application of the control command. Since the framework hides the protocols details about network communication it is always necessary to be sure that delays does not affect the packet relay.

This is true especially in multi-hop networks where the path of packets is not fixed but it is function of a series of parameters that cannot be monitored run-time.

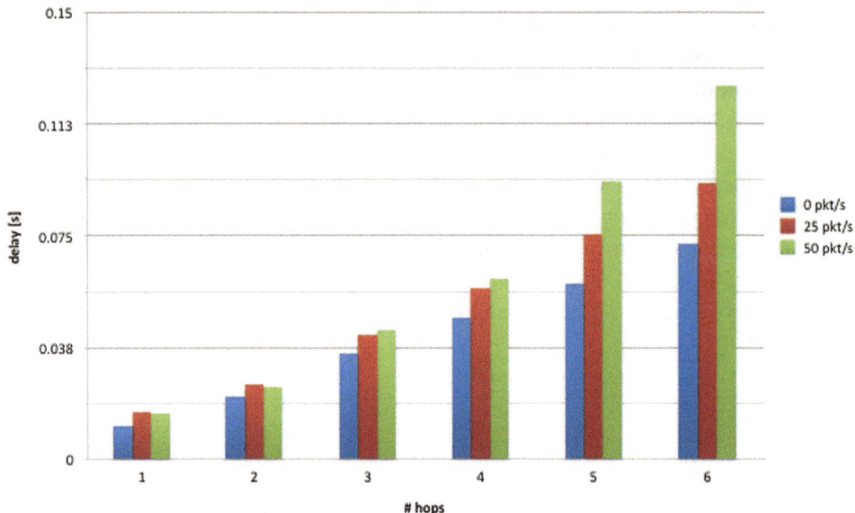

Fig. 4.10 Delay versus Hops in a multihop ZigBee network varying the traffic traversing each intermediate hops

In Fig. 4.10 is presented a graph representing the delay inside a ZigBee network in function of the number of hops traversed by a packet, using as parameter the outbound traffic (in packet/s) affecting each hop. The experimental setup is composed by ZigBee Ember EM250 nodes configured as routers. The path to follow within the network is hard-coded inside the payload of the packet itself while the node are subject to a high rate of packets sent towards the coordinator.

For a big network, a long path of 8 hops and with high traffic the delay is not above 130 ms. This value is not able to affect significantly the control considering that the resolution process of the LP problem is much more longer and therefore determines the throughput of the entire system.

SPINE2 introduces also a slightly delay in sending routines and data acquisition. If the sampling period of sensors is not too short the delay in reading sensors is negligible. In particular, the MPC can add some constrains on sampling rate to avoid a too frequent access to sensors. The delay in sending routines can be neglected because the additional average delay of 10 ms does not compromise the resolution of the linear programming problem which is more intense and time demanding.

4.9 Conclusions

This chapter presented several health-care research prototypes based on Body Sensor Network (BSN) technologies and programmed atop a domain-specific framework called SPINE.

BSNs represent an emerging technology that is potentially disruptive for the next generation of mobile health-care applications as it enables real-time, continuous and non-invasive monitoring of several vital signs of the assisted living without interfering with daily life activities.

We promoted the use of SPINE as middleware layer atop which to build mobile healthcare applications based on BSNs fulfilling requirements of effectiveness and efficiency. In particular, a physical activity recognition system, an emotional stress detector, a physical rehabilitation assistant, and a fall detector have been described and analysed. Furthermore, we introduced two basic applications patterns based on SPINE2 [48, 49] which are the building blocks of more complex mobile healthcare applications.

However, despite BSN potentials and the number, variety and quality of BSN-based health-care research prototypes, there is still a gap to fulfil before this technology can hit significantly the market with commercial solutions. Besides reducing the size of the wearable devices and improve the quality in terms of robustness and accuracy, a promising future outlook for this systems to be successfully commercialized is related to the integration between BSNs and cloud computing platforms to gain better processing and recording capabilities for data analysis e long-term storage [50].

References

1. Ko J, Lim JH, Chen Y, Musvaloiu-E R, Terzis A, Masson GM, Gao T, Destler W, Selavo L, Dutton RP (2010) Medisn: medical emergency detection in sensor networks. ACM Trans Embed Comput Syst 10(1):11:1–11:29. doi:10.1145/1814539.1814550
2. Virone G, Wood A, Selavo L, Cao Q, Fang L, Doan T, He Z, Stoleru R, Lin S, Stankovic JA (2006) An advanced wireless sensor network for health monitoring. In: Transdisciplinary conference on distributed diagnosis and home healthcare (D2H2), pp 2–5. http://www.cs.virginia.edu/papers/d2h206-health.pdf
3. Malan DJ, Welsh M, Smith MD (2008) Implementing public-key infrastructure for sensor networks. ACM Trans Sens Netw 4(4):1–23. doi:10.1145/1387663.1387668
4. Malan D, Fulford-Jones T, Welsh M, Moulton S (2004) Codeblue: an ad hoc sensor network infrastructure for emergency medical care. In: MobiSys 2004 workshop on applications of mobile embedded systems (WAMES 2004)
5. Zhang M, Sawchuk A (2009) A customizable framework of body area sensor network for rehabilitation. In: Proceedings of the 2nd international symposium on applied sciences in biomedical and communication technologies, ISABEL 2009, IEEE Press, pp 24–27
6. SunSPOT-website (2012) http://www.sunspotworld.com
7. Gravina R, Guerrieri A, Fortino G, Bellifemine F, Giannantonio R, Sgroi M (2008) Development of body sensor network applications using SPINE. In: Conference proceedings—IEEE international conference on systems, man and cybernetics, pp 2810–2815
8. Bellifemine F, Fortino G, Giannantonio R, Gravina R, Guerrieri A, Sgroi M (2011) SPINE: a domain-specific framework for rapid prototyping of WBSN applications. Softw Pract Experience 41(3):237–265
9. Gravina R, Alessandro A, Salmeri A, Buondonno L, Raveendranafhan N, Loseu V, Giannantonio R, Seto E, Fortino G (2010) Enabling multiple BSN applications using the spine framework. In: Proceedings of the 2010 international conference on body sensor, networks, pp 228–233

10. Buondonno L, Fortino G, Galzarano S, Giannantonio R, Giordano A, Gravina R, Guerrieri A (2009) Programming signal processing applications on heterogeneous wireless sensor platforms. In: Proceedings of the 5th IEEE international workshop on intelligent data acquisition and advanced computing systems: technology and applications, IDAACS'2009, pp 682–687
11. SPINE-website (2012) http://spine.deis.unical.it
12. Lombriser C, Roggen D, Stiger M, Trister G (2007) Titan: a tiny task network for dynamically reconfigurable heterogeneous sensor networks. In: 15 Fachtagung Kommunikation in Verteilten Systemen (KiVS) 15(127–138)
13. Aiello F, Fortino G, Gravina R, Guerrieri A (2011) A Java-based agent platform for programming wireless sensor networks. Comput J 54(3):439–454
14. Maurer U, Smailagic A, Siewiorek DP, Deisher M (2006) Activity recognition and monitoring using multiple sensors on different body positions. In: Proceedings of the international workshop on wearable and implantable body sensor networks, IEEE Computer Society, BSN 2006, pp 113–116
15. Dinh A, Teng D, Chen L, Shi Y, McCrosky C, Basran J, Del Bello-Hass V (2009) Implementation of a physical activity monitoring system for the elderly people with built-in vital sign and fall detection. In: 6th international conference on information technology: new generations, 2009, series ITNG '09, pp 1226–1231
16. Dinh A, Teng D, Chen L, Ko SB, Shi Y, McCrosky C, Basran J, Del Bello-Hass V (2009) A wearable device for physical activity monitoring with built-in heart rate variability. In: 3rd international conference on bioinformatics and biomedical engineering, 2009, ICBBE 2009, pp 1–4
17. Zappi P, Lombriser C, Stiefmeier T, Farella E, Roggen D, Benini L, Tröster G (2008) Activity recognition from on-body sensors: accuracy-power trade-off by dynamic sensor selection. In: Proceedings of the 5th European conference on wireless sensor networks, EWSN'08. Springer, Berlin, pp 17–33
18. Zhu C, Sheng W (2011) Wearable sensor-based hand gesture and daily activity recognition for robot-assisted living. IEEE Trans Syst Man Cybern Part A Syst Hum 41(3):569–573
19. Carnì D, Fortino G, Gravina R, Grimaldi D, Guerrieri A, Lamonaca F (2011) Continuous, real-time monitoring of assisted livings through wireless body sensor networks. In: Proceedings of the 6th IEEE international conference on intelligent data acquisition and advanced computing systems: technology and applications, IDAACS'2011, vol 2, pp 872–877
20. Raveendranathan N, Galzarano S, Loseu V, Gravina R, Giannantonio R, Sgroi M, Jafari R, Fortino G (2012) From modeling to implementation of virtual sensors in body sensor networks. IEEE Sens J 12(3):583–593
21. Panuccio P, Ghasemzadeh H, Fortino G, Jafari R (2011) Power-aware action recognition with optimal sensor selection: an adaboost driven distributed template matching approach. In: Proceedings of the 1st ACM workshop on mobile systems, applications, and services for healthCare—co-held with ACM SenSys 2011
22. Cover T, Hart P (1967) Nearest neighbor pattern classification. IEEE Trans Inf Theory 13:21–27
23. Pudil P, Novovicova J, Kittler J (1994) Floating search methods in feature selection. Pattern Recogn Lett 15(11):1119–1125
24. Food, Nutrition Board NAoS (2002) Dietary reference intakes for energy, carbohydrates, fiber, fat, protein and amino acids. National Academy Press, Washington
25. U.S. Department of Health & Human Services UD, for Disease Control & Prevention C, for Chronic Disease Prevention & Health Promotion TNC (1996) Physical activity and health: a report of the surgeon general
26. Chen K, Sun M (1997) Improving energy expenditure estimation by using a triaxial accelerometer. J Appl Physiol 83(6):2112–2122
27. Mizell D (2003) Using gravity to estimate accelerometer orientation. In: Proceedings of the 7th IEEE international symposium on wearable computers, IEEE Computer Society, ISWC'03, pp 252–253
28. Andreoli A, Gravina R, Giannantonio R, Pierleoni P, Fortino G (2010) SPINE-HRV: a BSN-based toolkit for heart rate variability analysis in the time-domain. Lecture notes in, electrical engineering, LNEE, vol 75, pp 369–389

29. Andreoli A, Gravina R, Giannantonio R, Pierleoni P, Fortino G (2010) Time-domain heart rate variability analysis with the spine-HRV toolkit. In: Proceedings of the 3rd international conference on pervasive technologies related to assistive environments
30. McEwen B (1998) Protective and damaging effects of stress mediators. N Engl J Med 338(3):171–179
31. Segerstrom SC, Miller GE (2004) Psychological stress and the human immune system: a meta-analytic study of 30 years of inquiry. Psychol Bull 130(4):601–630
32. Yang HK, Lee JW, Lee KH, Lee YJ, Kim KS, Choi HJ, Kim DJ (2008) Application for the wearable heart activity monitoring system: analysis of the autonomic function of HRV. In: Proceedings of the 30th annual international conference on engineering in medicine and biology society, EMBS 2008, IEEE Press, pp 1258–1261
33. Raya C, Torrent M, Parera J, Angulo C, Catala A (2007) KneeMeasurer: a wearable interface for joint angle measurements. In: International congress on domotics, robotics and remote-assistance for all
34. Bo APL, Hayashibe M, Poignet P (2011) Joint angle estimation in rehabilitation with inertial sensors and its integration with kinect. In: Annual international conference of the IEEE engineering in medicine and biology society
35. Cooper G, Sheret I, McMillian L, Siliverdis K, Sha N, Hodgins D, Kenney L, Howard D (2009) Inertial sensor-based knee flexion/extension angle estimation. J Biomech 42(16):2678–2685
36. Benocci M, Tacconi C, Farella E, Benini L, Chiari L, Vanzago L (2010) Accelerometer-based fall detection using optimized ZigBee data streaming. Microelectron J 41(11):703–710. doi:10.1016/j.mejo.2010.06.014, IEEE international workshop on advances in sensors and interfaces 2009
37. Chen J, Kwong K, Chang D, Luk J, Bajcsy R (2005) Wearable sensors for reliable fall detection. In: 27th annual international conference of the IEEE engineering in medicine and biology society 2005, IEEE-EMBS 2005, pp 3551–3554. doi:10.1109/IEMBS.2005.1617246
38. Kangas M, Konttila A, Lindgren P, Winblad I, Jamsa T (2008) Comparison of low-complexity fall detection algorithms for body attached accelerometers. Gait Posture 28(2):285–291. doi:10.1016/j.gaitpost.2008.01.003
39. Caione C, Brunelli D, Benini L (2012) Distributed compressive sampling for lifetime optimization in dense wireless sensor networks. IEEE Trans Industr Inf 8(1):30–40. doi:10.1109/TII.2011.2173500
40. Caione C, Brunelli D, Benini L (2010) Compressive sensing optimization over zigbee networks. In: 2010 international symposium on industrial embedded systems (SIES), pp 36–44. doi:10.1109/SIES.2010.5551380
41. Chiasserini C (2002) On the concept of distributed digital signal processing in wireless sensor networks. In: Proceedings of MILCOM 2002, vol 1, pp 260–264. doi:10.1109/MILCOM.2002.1180450
42. Nikolaou M (2001) Model predictive controllers: a critical synthesis of theory and industrial needs. Advances in chemical engineering (Elsevier) 26:131–204. doi:10.1016/S0065-2377(01)26003--7
43. Moser C, Thiele L, Brunelli D, Benini L (2010) Adaptive power management for environmentally powered systems. IEEE Trans Comput 59(4):478–491. doi:10.1109/TC.2009.158
44. Dondi D, Bertacchini A, Larcher L, Pavan P, Brunelli D, Benini L (2008) A solar energy harvesting circuit for low power applications. In: Sustainable energy technologies, 2008. ICSET 2008. IEEE international conference on, pp 945–949. doi:10.1109/ICSET.2008.4747143
45. Carli D, Brunelli D, Bertozzi D, Benini L (2010) A high-efficiency wind-flow energy harvester using micro turbine. In: Power electronics electrical drives automation and motion (SPEEDAM), 2010 international symposium on, pp 778–783. doi:10.1109/SPEEDAM.2010.5542121
46. Brunelli D, Dondi D, Bertacchini A, Larcher L, Pavan P, Benini L (2008) Photovoltaic scavenging systems: modeling and optimization. Microelectron J 40(9):1337–1344. doi:10.1016/j.mejo.2008.08.013

47. Porcarelli D, Brunelli D, Magno M, Benini L (2012) A multi-harvester architecture with hybrid storage devices and smart capabilities for low power systems. In: Power electronics, electrical drives, automation and motion (SPEEDAM), 2012 international symposium on, pp 946–951. doi:10.1109/SPEEDAM.2012.6264533
48. Fortino G, Guerrieri A, Bellifemine F, Giannantonio R (2009) Platform-independent development of collaborative wireless body sensor network applications: Spine2. In: Proceedings of the 2009 IEEE international conference on systems, man and cybernetics, pp 3144–3150
49. Fortino G, Guerrieri A, Bellifemine F, Giannantonio R (2009) Spine2: developing bsn applications on heterogeneous sensor nodes. In: Proceedings of the IEEE international symposium on industrial embedded systems, pp 128–131
50. Fortino G, Pathan M, Di Fatta G (2012) BodyCloud: integration of cloud computing and body sensor networks. In: Proceedings of the 4th IEEE international conference on cloud computing technology and science

Chapter 5
Markets for Cooperating Objects

Modern enterprises need to be agile and to dynamically support decision making processes at several levels. In order to be able to take efficient decisions and manage the resources in an optimal way, a direct link to the timely provision of information residing in all layers between the enterprise services and the resources needs to be established. This increases visibility at a very discrete level and can provide insights on how specific problems can be avoided or tackled. However monitoring is not enough, as controlling and adapting the behaviour of the resources needs to take place in order to close the loop [1].

Existing business processes may become more accurate since information taken directly from the point of action can be used to manage processes and related decision-making procedures. The continuous evolution of embedded and ubiquitous computing technologies, in terms of decreasing costs and increasing capabilities, may even lead to the distribution of existing business processes to the "network edges" and can overcome many limitations of existing centralized approaches. Cooperating Objects offer these capabilities by introducing cooperation as the key principle that may enhance future devices, systems and applications.

The domain of Cooperating Objects is still at its dawn; however its impact is estimated to be so broad and significant that could drastically change the future application and services. Numerous market analyses seem to point towards this direction also. It is important to understand that Cooperating Objects is a huge domain with applications in several fields [2], and therefore it is very difficult to set the limits and estimate its total value. As such we indicatively refer only to some markets that fall in the category of the Cooperating Objects such as the (wireless) sensors, networked embedded systems etc.

Contributors of this chapter include: Nils Aschenbruck, Jan Bauer, Davide Brunelli, Enrique Casado, Armando Walter Colombo, Gianluca Dini, Elisabetta Farella, Giancarlo Fortino, Christoph Fuchs, Roberta Giannantonio, Philipp Maria Glatz, Raffaele Gravina, Stamatis Karnouskos, Paulo Leitão, Pedro José Marrón, José Ramiro Martínez-de Dios, Simone Martini, Marco Mendes, Luis Merino, Daniel Minder, Luca Mottola, Amy L. Murphy, Aníbal Ollero, Lucia Pallottino, Gian Pietro Picco, José Pinto, and Thiemo Voigt.

S. Karnouskos et al., *Applications and Markets for Cooperating Objects*,
SpringerBriefs in Cooperating Objects,
DOI: 10.1007/978-3-642-45401-1_5, © The Author(s) 2014

5.1 Market Overview

Cooperating Objects are an integral part of the future Internet of Things. The latter is expected to enable unprecedented interconnection of networked embedded devices and further blur the line between the real and virtual world. If we take a closer look at individual domains, we will see a tremendous growth on the network side, and information will be provided by networked embedded devices. Several predictions are made about the status of things connected to the Internet, which forms also a big part of the basis for Cooperating Object approaches to flourish. Similarly high expectations are made on the Cooperating Object domains that could be impacted such as the Smart Grid, smart cities, industrial automation, aviation, robotics etc. some of which are already depicted in this book.

According to Håkan Djuphammar, VP of systems architecture at Ericsson, "[In 10 years' time], everything has connectivity. We're talking about 50 billion connections, all devices will have connectivity ..." [3]. This was reinforced by the Ericsson President and CEO Hans Vestberg who mentioned that 50 billion devices will be connected to the web by 2020. Intel's John Woodget, global director, Telecom sector has a more moderate prediction, in the range of 20 billion connected devices by 2020 [3].

According to the Broadband Commission for Digital Development [4], "worldwide, mobile phone subscriptions exceeded already the 6 billion in early 2012", and "by 2020, the number of connected devices may potentially outnumber connected people by six to one". In the same report the total networked devices is expected to reach the 25 billion by 2020.

According to Gartner's "Top 10 Strategic Technology Trends for 2013" [5] already "...over 50 % of Internet connections are things. In 2011, over 15 billion things on the Web, with 50+ billion intermittent connections. By 2020, over 30 billion connected things, with over 200 billion with intermittent connections. Key technologies here include embedded sensors, image recognition and NFC".

Getting down to the smartgrid specific statements, Marie Hattar, vice president of marketing in Cisco's network systems solutions group, estimated in 2009 that the smart-grid network will be "100 or 1,000 times larger than the Internet" [6]. Similarly Vishal Sikka, CTO of SAP, stated in 2009 that "The next billion SAP users will be smart meters" [7]. Only for installing smart meters in homes an estimated $4.8 billion will be spent according to ABI Research [8].

According to Pike Research the market for energy management systems (including Wireless Sensor Networks, lighting controls, heating and cooling management in buildings) will turn into a $6.8 billion a year market by 2020 and will generate investment of $67.6 billion between 2010 and 2020 [9]. They also note that a total of $4.3 billion will be spent on the installation, maintenance, and management services for smart grids by 2015 [10].

According to SESAR Joint Undertaken (www.sesarju.eu) the deployment of the future ATM net-centric infrastructure (which as demonstrated is a domain that Cooperating Object approaches are applied) will provide end-users with early availability of the most accurate information on weather situation, air congestion, situation on

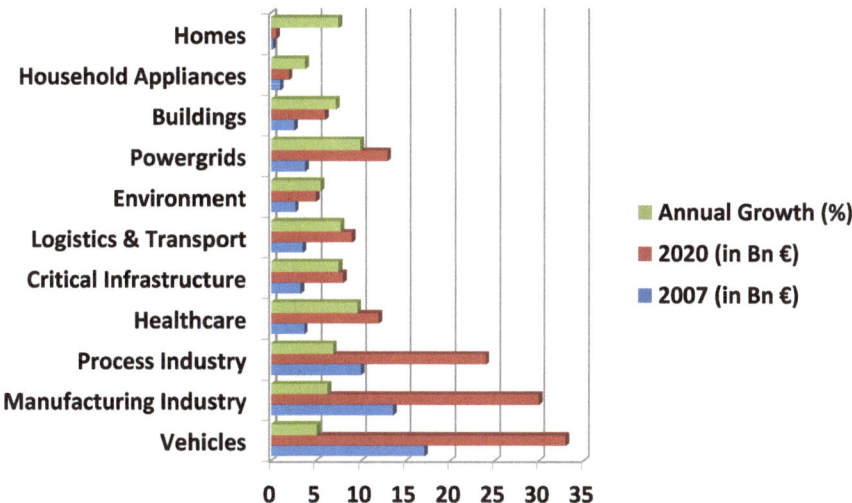

Fig. 5.1 Monitoring and control market 2007–2020 [13]

the ground, etc. This will help air traffic controllers and pilots avoid for example unnecessary reroutes, or taxiing on the ground. As a result, less fuel is burnt and resources used more efficiently with positive effects on the environment. It is estimated that by reducing taxi time by 10 %, the total of European annual savings to airlines is at 145.000 tons fuel with translates to €120 million or 475.000 tons of CO_2.

These are just some examples, depicting the fact that we are still at the dawn of a new era [11]. A \$4.5 trillion impact is estimated [12] by 2020 on people and businesses stemming from the sale of connected devices and services. There is a clear trend where networked embedded devices will blend with the everyday lives and directly or indirectly affect them. Cooperation among Cooperating Objects at local and system-wide level may create new business opportunities in the future.

The main focus of Cooperating Objects is in coupling the physical and virtual worlds; they do this via monitoring and control activities. The most important market sectors potentially affected by and from Cooperating Objects are depicted in Fig. 5.1. As reported in the European Commission study [13], the world M&C market is expected to grow from €188 billion in 2007, by €300 billion, reaching €500 billion in 2020. Between 2007 and 2020 the European monitoring and control domain is expected to grow at a 5.7 % rate annually. With a share of €61.5 billion today, Europe represents 32 % of this market. Services, with more than 50 % of the market value, have the biggest share. Together, three application markets: Vehicles, Manufacturing and Process industries represent 60 % of total monitoring and control market. Healthcare, critical infrastructures and logistic and transport follow closely. At the moment, the Home is still considered a small niche market.

The overall market where Cooperating Objects technologies are contributing is expected to grow significantly until 2020 (see Fig. 5.1). Hardware is expected to have a relative small growth due to decreasing prices; this does not hold true for network devices which will have an exponential growth in the next years. Services are expected to dominate the market i.e., next generation of products or components is included in service packages. This emergence of new services will create also the need for next generation products e.g., in environmental regulations, energy efficiency etc. The Total Cost of Ownership (TCO) is expected to be extended and include issues such as precision maintenance, asset management, production tools life extension with higher maintenance needs, more secure and safe installation and infrastructures.

Several innovations relevant to the Cooperating Object domain are expected. In Components, increasing computing power and integration, intelligent communicating local components, standardization and lower prices are foreseen. In networks, IP will be everywhere, networks will be transparent across application sectors, and service oriented approaches will be dominant. On the Services, it is expected that we will have a largely industrialized version of them. As most of the technologies are already in place, what remains is the optimal exploitation of them. Many technologies still seem futuristic and with prohibitive cost for mass-application usage. As such the evolution of the domain will not be heavily based on the technology as such only but directly linked to different business models that are connected to it.

5.2 Featured Applications

Table 5.1 provides an overview of the market relevance for the featured applications. We discuss them next in detail.

5.2.1 Deployment and Management of Cooperating Objects

Monitaring Railway Bridges: Monitoring bridges is currently one of the most costly activities in the domain of structural engineering. The difficulty of accessing the target sites, along with the required data accuracy, concur to increase development and installation costs. The commercially available systems, mostly based on wired technology, also require specialized training and personnel for the installation. Cooperating Objects technology may greatly reduce the relative costs, due to the ease of installation. The major hampering factor, however, is currently related to the short lifetime of wireless systems compared to their wired counterpart. The latter are potentially able to run for decades. We maintain that these applications of Cooperating Objects may eventually reach the market only when practical energy harvesting technology will be at disposal.

Cooperative Industrial Automation Systems: Web services on resource constrained devices are already available in one form or another (e.g., via the DPWS

Table 5.1 Envisioned market maturity for the featured applications

Market view ⇒ Applications ⇓	Market state	1–5 years	5–10 years	beyond 10 years	comments
Monitoring railway bridges	tesbed deployments for system testing	large scale prototype deployments	commercial availability		To reach the market, standardization, provable performance guarantees, and practical energy harvesting technology are required
Cooperative industrial automation systems	some prototypes are available		partial commercial availability	production systems and solutions deployed	Dependencies on migration of current systems. Industrial infrastructures include long-running (10-15 years) plans
Light-weight bird tracking sensor nodes	numerous deployments have been completed	ongoing deployments and further miniaturization			
Public safety scenarios	first products are available		better devices with better batteries should raise deployment rates.		Standardization is needed for future public safety systems to enable cross-boarder cooperation.
Road tunnel monitoring and control	prototype operational in high-traffic tunnel for last 2 years, European patent awarded	in commercial availability			System extensions are planned to expand system functionality
Mobility in industrial scenarios	some prototypes are available	partial commercial availability	production systems and solutions deployed		Centralized solutions are already available. Additional steps are necessary for the industrialization of decentralized solutions.

Continued

Table 5.1 Continued

Market view ⇒ Applications ⇓	Market state	1–5 years	5–10 years	beyond 10 years	comments
Mobility in civil security and protection	some prototypes are available	partial commercial availability	partial systems and solutions deployed	production systems and solutions deployed	Solutions based on WSN are available. The cooperation between WSN and UAS is still at research level. Regulations on UAS flight is still pending.
Physical activity recognition	some prototypes are available	partial commercial availability	production systems and solutions deployed		
Real-time physical energy expenditure	commercial availability	production systems and solutions deployed			
Emotional stress detection	partial commercial availability	partial commercial availability	production systems and solutions deployed		Many other emotions will be detected and these systems will be connected with the Smart cities environment
Physical rehabilitation	some prototypes are available		production systems and solutions deployed		
Energy aware fall detection	some prototypes are available	partial commercial availability	production systems and solutions deployed		
Distributed digital signal processing	needs extensive characterization before exploitation		partial commercial availability	production systems and solutions deployed	

Continued

Table 5.1 Continued

Model predictive control	needs extensive characterization before exploitation	partial commercial availability	partial commercial availability	production systems and solutions deployed
Mobility in ocean scenarios	some prototypes are available		partial systems and solutions deployed	production systems and solutions deployed
Person assistance in Urban scenarios	some prototypes are available		partial commercial availability	production systems and solutions deployed
Mobility in air traffic management	initial prototypes development		partial commercial availability	production systems and solutions deployed Future ATM Operational Concept that will rely on cooperative capabilities is planned to be initially operative in 2020 in EU and in 2025 in USA

or even simple REST services) by industry. Although most publicly known and demonstrated efforts were done mostly in cutting edge research projects, we see slowly the inclusion of devices in industrial infrastructures that among their traditional functionality offer also access over simple web services. It is expected that such devices will be mainstream in the mid-term (next 5–10 years), and hence in the future SOA-driven approaches will be a reality in the future factories as well as other critical infrastructures. The existence of such devices is just the first step towards creating more sophisticated applications and realize future cooperative industrial automation systems. The latter is only expected to become mainstream in the long-term (in 10+ years). A significant challenge is also the migration of the industrial infrastructures and the support at their lifecycle, since in many domains they are set with state-of-the art proven technology for the next 10–15 years, and not significant technology changes are done after that. However, Cooperative Industrial Automation is a longer term vision and for its realization that goes beyond simple interactions, significant research investment will need to be made, while real-world demonstrators should make sure that the technology is mature and reliable enough to enter live production systems.

Light-Weight Bird Tracking Sensor Nodes: UvA BiTS is targeting wireless tracking applications operating under challenging constraints. The main constraints being posed by the BiTS application that are solved are longevity (demanding for low-power operation and energy harvesting), mechanical endurance (birds might pick on the device) and dependability (state-recovery mechanisms), size and weight (down to 12 g). There is no other platform available today meeting those requirements. The applications that come closest to those requirements are for animal-tracking devices employing satellite communication modules. However, those devices are two orders of magnitude more expensive or even worse. Future perspectives for opening up a market can rely on case studies with many successful deployments proving the capability of the system. This is expected to overcome the commonly encountered hurdle of low trustworthiness of innovative platforms and applications not being considered possible or reliable before.

Public Safety Scenarios: First products in the mesh network domain targeting public safety scenarios are available on the market. However, for a broader deployment, lower costs and extended functionality concerning robustness and accuracy are crucial. In the research domain, robust and energy-aware protocols and algorithms for sensor and communication management are the important challenges. Overall, improved batteries, energy harvesting as well as more robust and sufficient hardware will further have a positive impact for the future.

Road Tunnel Monitoring and Control: The system described in Sect. 2.6 is the first of its kind to demonstrate that significant energy savings can be achieved by adapting light levels inside tunnels based on distributed sensing. This novelty has been confirmed by the awarding of a European patent in March 2012 for the overall architecture for adaptive lighting, of which the WSN is an integral element. Further, the ease of installation of the WSN technology means that the system can be easily deployed in existing tunnels without significant modifications to the infrastructure. This is particularly important in ageing tunnels where safety is a concern. The system had been implemented and is integrated with the industrial SCADA infrastructure developed by Siemens. Additional steps are necessary for the industrialization of the hardware, e.g., to ease the battery replacement and most likely to strengthen the connection between the mote and the sensor board.

5.2.2 Mobility of Cooperating Objects

Mobility in Industrial Scenarios: Some products in the mobility in industrial scenarios are already available on the market: they are mainly based on centralized architecture in which robots receive requests and tasks from a central base station and act basing on them without depending upon any other robot to accomplish its task. Additional steps are necessary for the industrialization of the hardware and software solutions implementing decentralized architectures, e.g., hardware reduction, to increase robustness and most likely to strengthen the connection between robots. In this sense, lower costs and more robust and reliable hardware will further have a positive impact for a future broader deployment of decentralized architectures.

Mobility in Air Traffic Management: The future ATM paradigm will represent a huge scenario of mobile (aircraft) and static (ground-based) Cooperating Objects. This new operational concept will make extensive use of a network infrastructure that will establish a proper connectivity among all stakeholders. This global infrastructure will be capable of providing accurate information with higher quality of service to any system connected to the network. Such infrastructure will include not only the physical layers required to transmit the data (datalinks and satellite or VHF communications), but also the upper layers which include the protocols and middleware technologies for enabling a proper and robust connectivity of all participants. The challenge will be to integrate not only the new designed on-board or on-ground subsystems, but also to do it with all legacy systems which had been adapted to be part of the community. Although the final aim is to have all Cooperating Objects integrated into the same network, there will be different deployment stages during which many diverse users equipped with a variety of capabilities will need to interoperate seamlessly without affecting safety.

Mobility in Ocean Scenarios: Cooperating Objects in ocean scenarios have a wide range of applications ranging from security to maintenance. The greater part of the market is in security applications, which are dominated by strong companies. Civil applications is currently a low-scale market but there are some commercial products including sensor nodes and robots, commercialized by small and medium enterprises. Civil market is more focussed on research than on applications. There are still technical issues such as energy availability, communication bandwidth and robustness of the hardware that constrain the use of this technology despite the large variety of potential applications. Security in communications and resilience to attacks are important research challenges. Improved batteries, communication systems as well as higher hardware robustness will further have a positive impact for the future.

Person Assistance in Urban Scenarios: This application can be seen as an extension of smart cities environments with mobile Cooperating Objects, like robots, which could assist persons in daily tasks like person and goods transportation, guiding, shopping, etc. For that, the mobile Cooperating Objects will cooperate with embedded infrastructure within the smart city, like camera and other sensor networks. Nowadays there are mainly prototypes from research projects. However, there are initial investments from private companies, like Google, which is investing project on self-driving cars [14] in cities (these autonomous cars have driven nowadays more than 150.000 km autonomously).

The development of social robotics, in which human-machine interaction plays a major role, will help with the introduction of these kind of applications in the market. One important aspect for the introduction of these kind of mobile Cooperating Objects are regulations regarding privacy [15].

Mobility in Civil Security and Protection: Some solutions based on static WSN are already available. There are commercial WSN nodes oriented towards different applications in civil security such as urban and industrial fires. The current state involving aerial Cooperating Objects such as Unmanned Aerial Systems (UAS) are still at research level. However, the results obtained have gained significant market interest. As an example, the results from the AWARE project (http://www.aware-project.net) originated products that were commercialized shortly after the end of the project. For instance, only four months after the completion of the project, the sales of FC III E SARAH (Electric Special Aerial Response Autonomous Helicopter), the new electrical helicopter developed by Flying-Cam within AWARE, were of 1.5 MEuro. The company expects sales of 62.5 MEuro by 2015. In the first four months after the end of the project this company obtained 1 MEuro selling aerial filming services using the helicopter. Some of the services were related to inspection of industrial infrastructures. They estimate that inspection tasks using their helicopter cost an average of 60 times less than traditional inspection.

Other companies developed systems for fire fighters protection using the results obtained from AWARE: WSN methods for real-time monitoring of pollutants. The company launched a plan (approx. 8.4 MEuro) to integrate the methods developed in AWARE in commercial products such as fire fighter costumes and fire trucks. This company is also developing products of real-time WSN monitoring of pollutants in tunnels and refineries.

5.2.3 Cooperating Objects in Healthcare Applications

We have taken a look at the following applications that fall within this category:

- Physical Activity Recognition
- Real-time Physical Energy Expenditure
- Emotional Stress Detection
- Physical Rehabilitation

- Energy Aware Fall detection
- Distributed Digital Signal Processing
- Model Predictive Control

WSNs have emerged in the recent years as one of the enabling technologies for e-Health services, both as body worn sensors and as environmental assistance networks. However, their practical application is still a challenge because of stringent requirements in terms of reliability, quality of service, privacy and security. For instance, end-to-end system *reliability* is a complex goal to achieve as these systems will be often used even at home by patients with motor disabilities and medical staff with little training; therefore, loss in quality due to operator misuse is a big concern.

An interesting tendency that is worth to be further investigated is the use of the *closed-loop approach*. In fact, the use of a loop (acquisition–processing–*actuation* and again acquisition ...) limits the need for continuous assistance by clinicians. Active guidance by use of *multiple kinds of feedback* (tactile, audio or visual) is welcome for user sensory augmentation, developing actuators to this purpose in the network, and hence providing steps forward far beyond the state of the art.

To reach both the goal of an independent system, self-calibrating and tuning, and the goal of an energy efficient system, *on-board processing* is one important technical challenge to be addressed, in order to delegate less and less tasks to external intervention, and to augment the transparency of the system from the user. The smart distribution of the processing can be strategic to limit data transmission and therefore power consumption due to wireless communication. Being the intelligence distributed, each component is smart and already offering important information (e.g., an accelerometer node can also offer results from tilt extraction, activity recognition, statistics on user quantity of movement per day, etc.). As a consequence also the entire system can fuse information at a higher abstraction level to understand user behaviours, calculate user training performance, regulate the overall behaviour of the system, care for the remote interfacing, etc., that is resulting in a *multi-functional system*.

The *microcontroller* marketplace has traditionally been heavily fragmented (8, 16 and 32-bit markets). However, with the introduction of innovative devices based on common architectures, such as ARM, the market is rapidly changing with ease of use, excellent code density, competitive pricing (32-bit devices are now widely available below \$2) and market consolidation all now becoming associated with high performance 32-bit machines.

The growth in 32-bit devices is being driven by a different set of market forces, including the need to improve code reuse across projects, application complexity, device aggregation, and system connectivity. Code reuse and the chance to develop with higher level languages, means the ability to guarantee *continuity* of a product and *availability of a service in time*. But also, it means that the *same platform can be used in multiple applications*, thanks to re-programming, *going beyond "one-sensor solution one disease"*.

Wireless technologies for improved energy efficiency are another key enabler. The most logical candidates for health applications are Zigbee and Bluetooth in their more recent versions, which are designed for low data rate, low power consumption, and low implementation costs targeted to applications such as the healthcare ones. These characteristics also comply with the guidelines provided by the Continua Health Alliance (an open industry consortium), whose mission is to establish an ecosystem of interoperable personal health systems in the healthcare.

The area of Micro/Nano System and Smart System Integration has an important and emerging role in many application fields, since it focuses on forming systems out of single components devoted to capturing information from the environment by using sensors, processing it by use of smart algorithms and filters, sending it to other processing unit or remote users and performing adequate action as a consequence of the knowledge extracted from it. This pipeline of operations is at the base of many different applications, from healthcare and rehabilitation to surveillance and smart buildings. While the processing capabilities of Integrated Circuits are virtually unlimited, this must be combined with the need for lifetime prolongation, miniaturization, low-cost and capabilities to interface with the external world while guaranteeing mobility to the user.

These specifications are required in *biomedical devices, which must be based on technology more and more autonomous and smart*. Nevertheless, such requirements cannot be easily reached at the same time. Energy for example, in biomedical electronics, is highly limited and size is not optimized if the priority is given on processing and interface with the external world, both in terms of communication and electrodes. A medical device powered by a low-power general purpose processor consumes approximately 10 mW. Current battery technology would accommodate approximately 3 days of operation. Alternatively, dedicated solutions, employing specialized low-power design techniques, consume approximately 8 μW, achieving more than 10 years of operation with the same battery. However, the kind of processing and interfacing allowed is limited [16].

Therefore, there is a dichotomy between highly miniaturized and ultra low-power systems, with basic processing and very limited communication capabilities and, on the other side, extremely smart devices, with rich interfacing capabilities, but not optimized in form-factor and limited lifetime or big batteries.

Recently, technical progress has made possible the realization of miniature kinematic sensors such as accelerometers and angular rate sensors with integrated conditioning and calibrating module. In addition, due to their very low consumption, these sensors can be battery powered and are promising tools for long-term ambulatory monitoring. In recent times, many studies have described custom-designed accelerometer-based systems, and task-specific data analysis techniques capable of detecting features of the walking pattern.

Enabling technologies: State of the art technology in biomedical wearable devices for biosignal monitoring consist in multi-sensorised devices where the main building blocks are a sensing/actuation unit, a power supply unit, a processing subsystem including analog-to-digital conversion and signal processing, and finally a communication subsystems. Each of these subsystems must be designed for minimum energy consumption. Specific design techniques such as aggressive voltage scaling, dynamic power-performance management and energy efficient signalling,

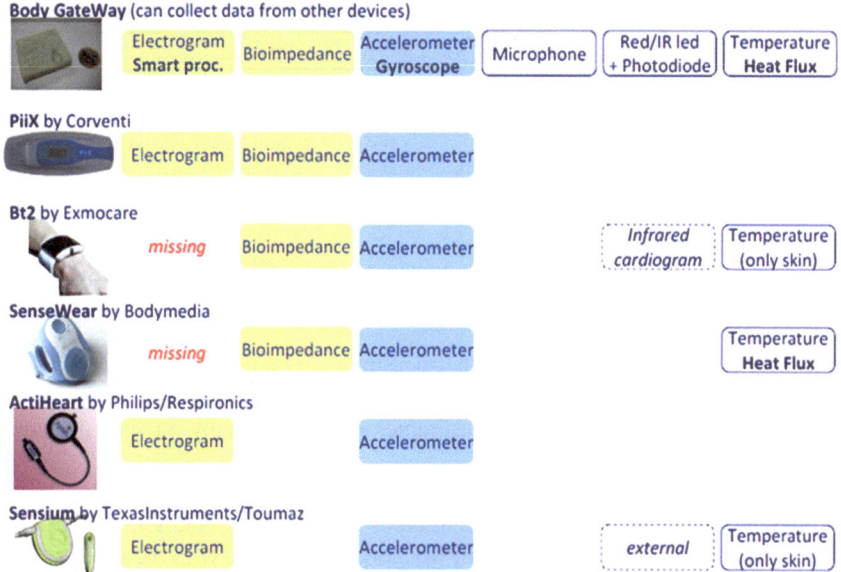

Fig. 5.2   Examples of monitoring devices using 32-bit processing wireless architecture

must be employed to adhere to the stringent energy constraint. Due to the impact in healthcare of smart systems, there are several examples of monitoring devices responding to the mentioned architecture.

Sensor technologies: Real-time data recording and processing of multi-physiological signals is yet a reality thanks to advances in new materials and signal processing research. A wide selection of medical sensors are currently available (from conventional sensors based on piezoelectric materials for pressure measurements to infrared sensors for body temperature estimation and optoelectronic sensors monitoring SpO_2, heart rate, and blood pressure). MEMS technologies have revolutionized many sectors of manufacturing, introducing low-cost accurate sensors for a wide range of measurements (e.g., microphones, inertial sensors, proximity sensors, CMOS elements). It is a current trend to provide not only the sensor itself (the transducer) but also a rich electronics implementing signal conditioning, analog-to-digital conversion and in many cases advanced filtering and buffering. In some cases a wireless communication system is included (e.g., ST MotionBee). The following table shows advantages of using MEMS, taking as example the comparison between a piezoelectric sensor and an equivalent MEMS device. It is worth noting that the advantage of MEMS is that the electronics is already included in the sensor with consequent lower overall costs, size and higher resolution.

Major challenges in sensor technologies are related to the target sensors and/or high quality electrodes that are required to be low-power, low-cost, small and at the same time maintain noise under low thresholds. Moreover, the choice of electrodes has a high impact on comfort and usability. Unfortunately, often choices for

Table 5.2 Advantage of using MEMS

Sensor (accelerometer)	Piezo	MEMS
Electronic output	Analog	Digital
Resolution	10 mV/g 1 mg	1 V/g
Size	1,000 mm^3 (without electronics)	10 mm^3
Power consumption	100 mW	3 mW
Price	200	10
Price of electronics for signal conditioning	50	0
Total price	250	10

improving accuracy and quality of the measurements are opposite to choices for augmenting user comfort. For example, larger electrodes or multiple electrodes in ECG guarantee better quality of measurements, but can annoy patients that must wear them for long-term monitoring.

Present research efforts are exploring alternative materials. Modern ECG monitoring devices use a variety of different types of ECG electrode. Examples of the different types include combinations of wet gel and hydro-gel with Ag and Ag/AgCl inks. Fluctuations in the electrode and skin interface potential and impedance but also user motion and environmental interferences deeply impact quality of measurements and artefacts presence. Currently, the most effective way to combat electrical changes due to fluctuations in the electrode and skin interface properties is through prevention—preparation of the skin to reduce its impedance and piezoelectric contribution to the signal and also through the reduction in overall movement of the subject for the duration of the monitoring. Current systems employ a number of methods for the reduction in resultant artefact signals.

A novelty in this field is the choice of Ag-AgCl pigments for screen printing on flexible polyester sheets. The interface can be reduced by special formulated ink to binder ratios; the thin hydrogel (doped with NaCl) and the levels of surface roughness at the electrode gel interface can cooperate to minimise impedance and along with the thinness of the ink to have very little EMF absorbed. This results in low motion artefact electrodes, without paying the price of complexity and increased size. More work can be done to optimise this process and better understand interfaces via multi-frequency impedance spectroscopy. For example the use of nano-based silver formulations or CNT formulations, screen printable gels, nano-structured interfaces and graphite tracking are an interesting research challenge to be explored.

5.3 The CONET Survey

A survey was carried out by CONET among selected experts from within and outside of the consortium. Several domains are expected to significantly benefit from Cooperating Objects. We have found out (depicted in Fig. 5.3) that especially monitoring

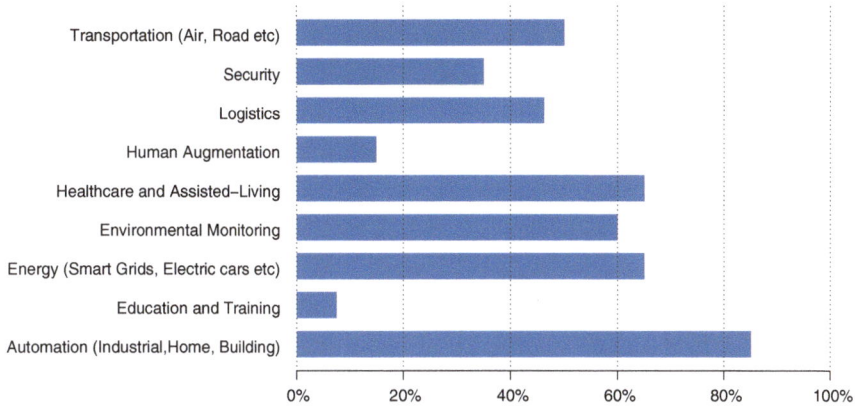

Fig. 5.3 Survey: Beneficiary domains

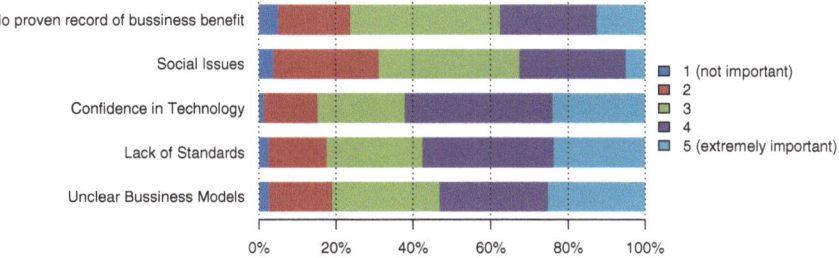

Fig. 5.4 Survey: Roadblock impact

and management in automation, energy, health and environment followed closely by the transportation, logistics, and security domain will be the major beneficiaries.

As it can be seen, the Cooperating Objects are expected to have significant impact on several domains. If we correlate Figs. 5.3 and 5.1 we can see that the emerging domains with the highest annual growth rate are the ones that may also benefit most from the success of Cooperating Objects. For the wide-spread adoption of Cooperating Objects technologies in mass-market products, several roadblocks are also identified (depicted in Fig. 5.4).

Confidence in technology is the most critical issue to be solved, closely followed by the lack of standards and unclear business models. Furthermore social issues and no proven record of business benefit are seen as having a moderate effect on the success of Cooperating Objects. Especially the last one is typical in the technology domain as the advances and benefits can not be fully envisioned nor widely understood. Although market predictions for the deployment and use of Cooperating Object relevant technologies is promising, the identified roadblocks will need to be tackled effectively if Cooperating Objects are to succeed. Additional details on the roadmap and the survey are described in [2, 17].

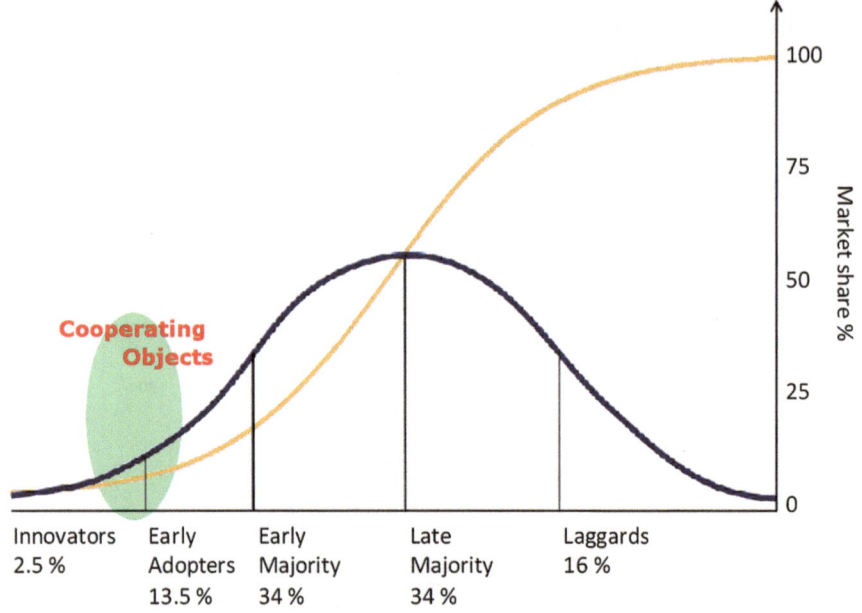

Fig. 5.5 Cooperating Objects in technology adoption lifecycle

5.4 Conclusion

The majority of the market growth predictions are constantly modified to the current business climate, therefore the aforementioned numbers should be taken "cum grano salis" and only as an indicative trend depicting the underlying potential; the future will tell if and at what timeline they will be validated.

Nevertheless, it is clear that there is promising potential in versatile domains, which could greatly benefit with the introduction of Cooperating Object technologies, ranging from automation (home, industrial, building) to healthcare, energy etc. We estimate that we are still in the dawn of a new era, and in the early phases of Rogers' technology adoption lifecycle as depicted in Fig. 5.5. We expect that the Cooperating Objects market will be cross-domain and strongly embedded in the fabric of success of other domains.

The impact of Cooperating Objects may affect many traditional industries and create significant business opportunities for companies across industries by opening up new markets and therefore may become an important factor of tomorrow's business environment and service-based economy. Its development holds the potential to provide market stakeholders with a competitive advantage in global markets, be it in terms of technologies or new services and applications. Finally, we consider that Cooperating Objects will act as an enabler for a wide range of applications and services, and hold the potential to empower sophisticated highly dynamic complex systems and applications in the long-term.

References

1. Karnouskos S (2009) Efficient sensor data inclusion in enterprise services. Datenbank-Spektrum 9(28):5–10
2. Marrón PJ, Minder D, Karnouskos S (2012) The emerging domain of cooperating objects: definition and concepts. Springer, Berlin. doi:10.1007/978-3-642-28469-4
3. Lomas N (2009) Online gizmos could top 50 billion in 2020. http://www.businessweek.com/globalbiz/content/jun2009/gb20090629_492027.htm
4. Biggs P (2012) The state of broadband 2012: achieving digital inclusion for all. Technical Report. The Broadband Commission for Digital Development. http://www.broadbandcommission.org/Documents/bb-annualreport2012.pdf
5. Savitz E (2012) Gartner: top 10 strategic technology trends for 2013. http://www.forbes.com/sites/ericsavitz/2012/10/22/gartner-10-critical-tech-trends-for-the-next-five-years/
6. LaMonica M (2009) Cisco: smart grid will eclipse size of internet. Interview. http://news.cnet.com/8301-11128_3-10241102-54.html
7. Mirchandani V (2009) The next billion SAP users will be smart meters. Interview. http://dealarchitect.typepad.com/deal_architect/2009/07/the-next-billion-sap-users-will-be-smart-meters.html
8. ABI Research (2010) Smart grid spending will top $45 billion by 2015. http://www.abiresearch.com/eblasts/archives/analystinsider_template.jsp?id=229
9. Pike Research (2009) Energy management systems for commercial buildings will garner $67 billion in investment by 2020. Press Release. http://www.pikeresearch.com/newsroom/energy-management-systems-for-commercial-buildings-will-garner-67-billion-in-investment-by-2020
10. Pike Research (2010) Smart grid managed services market to grow 75 % year-over-year between 2010 and 2011. Press Release. http://www.pikeresearch.com/newsroom/smart-grid-managed-services-market-to-grow-75-year-over-year-between-2010-and-2011
11. Karnouskos S, Marron PJ, Minder D (2013) Cooperating Objects Design Space and Markets, 4th International Workshop on Networks of Cooperating Objects for Smart Cities 2013 (CONET/UBICITEC 2013), Philadelphia, USA. http://ceur-ws.org/Vol-1002/paper4.pdf
12. Machina Research (2012) The connected life: a usd4.5 trillion global impact in 2020. http://connectedlife.gsma.com/wp-content/uploads/2012/02/Global_Impact_2012.pdf
13. European Commission (2008) Monitoring and control: today's market, its evolution till 2020 and the impact of ICT on these. Workshop Presentation. http://www.decision.eu/smart/SMART_9Oct_v2.pdf
14. Campbell M, Egerstedt M, How JP, Murray RM (2010) Autonomous driving in urban environments: approaches, lessons and challenges. Philos Trans R Soc A: Math Phys Eng Sci 368(1928):4649–4672. doi:10.1098/rsta.2010.0110
15. Sanfeliu A, Llacer MR, Gramunt MD, Punsola A, Yoshimura Y (2010)Influence of the privacy issue in the deployment and design of networking robots in european urban areas. Adv Robot 24:1873–1899
16. Chandrakasan AP, Verma N, Daly DC (2008) Ultralow-power electronics for biomedical applications. Annu Rev Biomed Eng 10:247–274
17. Marrón PJ, Karnouskos S, Minder D, Ollero A (eds) (2011) The emerging domain of cooperating objects. Springer, Berlin. doi:10.1007/978-3-642-16946-5

Chapter 6
Conclusions

Cooperating Objects build upon the amalgamation of the physical and virtual (business) world in order to provide some added value for future applications and services relying on both worlds. As we are moving towards a Trillion Node Network Infrastructure, where devices will be interconnected and cooperate, providing and consuming information available, collaborative and emergent behaviours are expected to appear that empower new innovative approaches. The vision of Cooperating Objects is to tackle the emerging complexity by cooperation and modularity. Achieving enhanced system intelligence by cooperation of smart embedded devices pursuing common goals is relevant in many types of perception and system environments.

The emerging domain of Cooperating Objects, is a very dynamic one that has the potential of drastically changing the way people interact with the physical world as well as how business systems integrate it in their processes. We are still at the dawn of an era, where a new breed of applications and services, strongly coupled with our everyday environment will revolutionize our lives even in a deeper way than the Internet has done in these past years.

Seamless cooperation and collaboration is necessary to realize an environment where the user services are provided in a distraction free manner. Traditional models support the cooperation either by providing peer-to-peer communication between devices or by utilizing an infrastructure. We believe that combining both of these approaches provides several advantages.

The domain of Cooperating Objects is still at its dawn; however, its impact is estimated to be so broad and significant that could change drastically the future applications and services. Numerous market analyses also point out this direction. It is important to understand that Cooperating Objects is a huge domain with applications spawning several fields, and, therefore, it is very difficult to set the limits and estimate its total value. We hope that within the pages of this book you have gotten a glimpse of research aspects, emerging applications and research challenges.

Contributors of this chapter include: Nils Aschenbruck, Jan Bauer, Armando Walter Colombo, Christoph Fuchs, Philipp Maria Glatz, Stamatis Karnouskos, Paulo Leitão, Marco Mendes, Luca Mottola, Amy L. Murphy, Gian Pietro Picco, and Thiemo Voigt.

S. Karnouskos et al., *Applications and Markets for Cooperating Objects*,
SpringerBriefs in Cooperating Objects,
DOI: 10.1007/978-3-642-45401-1_6, © The Author(s) 2014

Index